SpringerBriefs in Optimization

SpringerBriefs in Optimization showcase algorithmic and theoretical techniques, case studies, and applications within the broad-based field of optimization. Manuscripts related to the ever-growing applications of optimization in applied mathematics, engineering, medicine, economics, and other applied sciences are encouraged.

More information about this series at http://www.springer.com/series/8918

Neculai Andrei

A Derivative-free Two Level Random Search Method for Unconstrained Optimization

 Springer

Neculai Andrei
Center for Advanced Modeling and Optimization
Academy of Romanian Scientists
Bucharest, Romania

ISSN 2190-8354   ISSN 2191-575X   (electronic)
SpringerBriefs in Optimization
ISBN 978-3-030-68516-4   ISBN 978-3-030-68517-1   (eBook)
https://doi.org/10.1007/978-3-030-68517-1

Mathematics Subject Classification: 65K05; 90C30; 90C56; 90C90

This Springer imprint is published by the registered company Springer Nature Switzerland AG
The registered company address is: Gewerbestrasse 11, 6330 Cham, Switzerland

Preface

In many optimization problems from different areas of activity, the functions defining their mathematical model often show that their derivative information, mainly the gradient and the Hessian, is not available. For such problems, the derivative-free optimization methods are recommended. Although the derivative-free optimization methods have a long history, they are still quite active in science, and industry, having plenty of approaches and developments to optimize increasingly complex and diverse problems.

This book presents a new derivative-free optimization method based on random search, which is completely different from other derivative-free optimization methods. In it, the minimizing function is supposed to be continuous, lower bounded and its minimum value is known. The purpose is to determine the point where this minimum value is attained. The idea of this method is to catch a deep view of the landscape of the minimizing function around the initial point by randomly generating some trial points at two levels. Roughly speaking our method is close to pure random search, but extended at two levels. At the first level, a number of trial points are randomly generated in a domain around the initial point. The set of all trial points is called a complex. At the second level, in a domain around each trial point from the first level, the algorithm generates another number of trial points, the so-called local trial points. The number of the trial points, the number of the local trial points and the domains defined by their bounds in which these trial points are generated are specified by the user. The minimizing function is evaluated in all these points. At each iteration, the algorithm consists of a number of rules for replacing the trial points with local trial points and for generating new trial points where the function value is smaller. The algorithm uses some criteria for reducing the bounds defining the domains of the trial points, or the bounds of the domains of the local trial points, as well as criteria for the reduction of the complex defined by the trial points. A key element of the algorithm is the evolution of the maximum distance among the trial points and the evolution of the maximum distance among the trial points and the local trial points. It has been proved that these distances tend to zero, thus ensuring the convergence of the algorithm.

The algorithm has two phases. In the first one, the reduction phase, for a relatively small number of iterations, the algorithm produces an abrupt reduction of the function values. In this phase, the evolution of the function values presents some plateaus where the reduction of the function values is not significant. These plateaus correspond to those iterations where the bounds of the domains in which the trial points are generated are reduced (halved), or the complex is reduced. The second phase, the stalling phase, is characterized by a large number of iterations where the function values converge very slowly to a minimum value of the minimizing function.

To find a point x^* where the minimizing function f has a minimum value around an initial point, the algorithm generates a sequence of points $\{x_k\}$ for which the corresponding sequence of values $\{f(x_k)\}$ is monotonously decreasing and bounded. Therefore, the sequence $\{f(x_k)\}$ is convergent to a value $f(x^*)$. By continuity, the corresponding sequence $\{x_k\}$ is convergent to a point x^* for which $f(x^*) \leq f(x_0)$. Since there is no access to the gradient or to the Hessian of the minimizing function f, nothing can be said about the optimality of x^*, but this is common for all derivative-free optimization algorithms.

The numerical performances of the algorithm are reported for solving 140 unconstrained optimization problems, out of which 16 are real applications with the number of variables in the range [2, 50]. Comparisons with the Nelder-Mead algorithm show that our two-level random search method for the unconstrained optimization algorithm is more efficient and more robust. Finally, the performances of the algorithm for solving a number of 30 large-scale unconstrained optimization problems up to 500 variables are presented.

There are a number of *open problems* which refer to the following: selection of the number of trial or of the number of local trial points, selection of the bounds of the domains where the trial points and the local trial points are randomly generated and of a criterion for initiating the line search. All these points need further investigation. However, intensive numerical tests have shown that this approach based on the derivative-free two-level random search method for unconstrained optimization is able to solve a large diversity of problems with different structures and complexities.

This monograph is intended for graduate students and researchers in the field of mathematics, computer science, and operational research who work in optimization, particularly focused on derivative-free optimization, as well as for the industry bodies interested in solving unconstrained optimization problems for which the derivatives of the function defining the problem are impractical to obtain, unavailable and unreliable.

I express my gratitude to the **Alexander von Humboldt Foundation** for its generous support during the 2+ research years which I spent at several universities in Germany. I also owe Elizabeth Loew a great debt of gratitude for her encouragement and assistance in my efforts to carry out the project of this book during the pandemic. Last but not least, I want to thank my wife Mihaela for her precious help and competent advice.

Bucharest, Romania   Neculai Andrei
November 2020

Contents

List of Figures

List of Tables

Chapter 1
Introduction

Abstract The unconstrained optimization problem is presented: the derivative and the derivative-free methods, as well as the optimality conditions are discussed. A short presentation of the two-level random search method for the unconstrained optimization of functions, for which the derivative information is not known, is given. Some open problems associated to this method are also shown. It is emphasized that for derivative-free methods like the one developed in this book, the only thing we can obtain is a point, where the minimizing function value is smaller or equal to the value in its initial point. Nothing can be said about its optimality, but having in view the scarcity of information on the minimizing function, mainly derivative, the result obtained may be of use for practical considerations.

Keywords Unconstrained optimization · Derivative searching methods · Derivative-free searching methods · Optimality conditions

1.1 The Problem

The following unconstrained optimization problem

$$\min_{x \in \mathbb{R}^n} f(x) \tag{1.1}$$

is considered, where $f : \mathbb{R}^n \to \mathbb{R}$ is a real-valued function of n variables. The interest is in finding a *local minimizer* of this function, that is, a point x^*, so that

$$f(x^*) \leq f(x) \quad \text{for all } x \text{ near } x^*. \tag{1.2}$$

If $f(x^*) < f(x)$, for all x near x^*, then x^* is called a *strict local minimizer* of function f. Often, f is referred to as the *objective function*, while $f(x^*)$ as the *minimum* or the *minimum value*.

N. Andrei, *A Derivative-free Two Level Random Search Method for Unconstrained Optimization*, SpringerBriefs in Optimization,
https://doi.org/10.1007/978-3-030-68517-1_1

The local minimization problem is different from the *global minimization problem,* where a global minimizer, that is, a point x^* so that

$$f(x^*) \leq f(x) \quad \text{for all } x \in \mathbb{R}^n \tag{1.3}$$

is sought. This book only deals with the local minimization problems.

According to the properties of function f, the methods for solving (1.1) are classified as methods based on its differentiability, that is, methods using the gradient $\nabla f(x)$ and the Hessian $\nabla^2 f(x)$ of function f (or an approximation of it), and methods based only on the values of the minimizing function f computed at a sequence of points generated by using different rules. The methods from the first class are called *derivative optimization methods*, while those from the second class are called the *derivative-free optimization methods*. It is worth mentioning that there are some in-between methods known as *alternative for derivative-free optimization methods.* These are the *algorithmic differentiation* and the *numerical differentiation.*

1.2   Derivative Methods for Unconstrained Optimization

For solving (1.1), the derivative optimization methods implement one of the following two strategies: the *line search* and the *trust region*. Both these strategies are used for solving the real applications of unconstrained optimization (1.1). These strategies are described, for example, in: Gill et al. 1981; Luenberger 1984; Bazaraa et al. 1993; Bertsekas 1999; Conn et al. 2000; Nocedal and Wright 2006; Sun and Yuan 2006; Bartholomew-Biggs 2008; Andrei 1999, 2009, 2015; Luenberger and Ye 2016.

In the *line search strategy*, the corresponding algorithm chooses a direction d_k and searches along this direction from the current iterate x_k for a new iterate with a lower function value. Specifically, starting with an initial point x_0, the iterations are generated as:

$$x_{k+1} = x_k + \alpha_k d_k, \quad k = 0, 1, \dots, \tag{1.4}$$

where $d_k \in \mathbb{R}^n$ is the *search direction,* along which the values of function f are reduced, and $\alpha_k \in \mathbb{R}$ is the *step size* determined by a line search procedure. The main requirement is that the search direction d_k, at iteration k should be a *descent direction.* The algebraic characterization of descent directions is that:

$$d_k^T g_k < 0, \tag{1.5}$$

which is a very important criterion concerning the effectiveness of an algorithm. In (1.5), $g_k = \nabla f(x_k)$ is the gradient of f in point x_k. In order to guarantee the global convergence of some algorithms, sometimes it is required that the search direction d_k should satisfy the *sufficient descent* condition:

$$g_k^T d_k \leq -c\|g_k\|^2, \tag{1.6}$$

where c is a positive constant.

The most important line search methods are: the steepest descent method (Cauchy 1847), the Newton method, the quasi-Newton method (Davidon 1959; Broyden 1970; Fletcher 1970; Goldfarb 1970; Shanno 1970; Powell 1970), the limited-memory quasi-Newton method (Nocedal 1980), the truncated Newton method (Dembo et al. 1982; Dembo and Steihaug 1983; Deuflhard 1990), and the conjugate gradient method (Andrei 2020a).

In the *trust-region strategy*, the idea is to use the information gathered about the minimizing function f in order to construct a model function m_k whose behavior near the current point x_k is similar to that of the actual objective function f. In other words, the step p is determined by approximately solving the following subproblem:

$$\min_p m_k(x_k + p), \tag{1.7}$$

where the point $x_k + p$ lies inside the *trust region*. If the step p does not produce a sufficient reduction of the function values, then it follows that the trust region is too large. In this case, the trust region is reduced and the model m_k in (1.7) is re-solved. Usually, the trust region is a ball defined by $\|p\|_2 \leq \Delta$, where the scalar Δ is known as the *trust-region radius*. Of course, elliptical and box-shaped trust regions may be used.

Usually, the model m_k in (1.7) is defined as a quadratic approximation of the minimizing function f:

$$m_k(x_k + p) = f(x_k) + p^T \nabla f(x_k) + \frac{1}{2} p^T B_k p, \tag{1.8}$$

where B_k is either the Hessian $\nabla^2 f(x_k)$ or an approximation to it. Observe that each time when the size of the trust region, that is, the trust-region radius, is reduced after a failure of the current iterate, then the step from x_k to the new point will be shorter and usually will point to a different direction from the previous point.

By comparison, the line search and the trust region are different in the order in which they choose the *search direction* and the *step size* to move to the next iterate. The line search starts with a direction d_k and then determines an appropriate distance along this direction, namely the step size α_k. In the trust region, firstly the maximum distance is chosen, that is, the trust-region radius Δ_k, and then a direction and a step p_k that determine the best improvement of the function values subject to this distance constraint are computed. If this step is not satisfactory, then the distance measure Δ_k is reduced and the process is repeated.

A relatively new approach is the *p*-regularized methods, where the trust-region ball $\{x : \|x\|^2 \leq \Delta\}$ is replaced by a higher-order regularization term weighted by a positive parameter σ. Specifically, the *p-regularized subproblem* is the following unconstrained minimization problem:

$$\min_{x \in \mathbb{R}^n} \left\{ h(x) \equiv c^T x + \frac{1}{2} x^T B x + \frac{\sigma}{p} \|x\|^p \right\}, \tag{1.9}$$

where $p > 2$ is an integer, and $\sigma > 0$ is the *regularization parameter*. The regularization term $(\sigma/p)\|x\|^p$ in (1.9) determines that $h(x)$ is a coercive function, that is, $\lim_{\|x\| \to +\infty} h(x) = +\infty$, that is, (1.9) can always attain its global minimum even for nonpositive definite matrices B. Therefore, a local approximation (1.9) of f is constructed and solved. If the global minimizer of (1.9) gives a reduction in the value of f, then it is accepted, otherwise σ is increased in order to force the regularization. The most common choice to regularize the quadratic approximation of function f is the *p*-regularized subproblem (1.9) with $p = 3$, known as the *cubic regularization*. The cubic regularization was introduced by Griewank (1981) in order to develop an algorithm based on the Newton method that is affine-invariant and convergent to second-order critical points. (A second-order critical point of f is a point $\bar{x} \in \mathbb{R}^n$ satisfying $\nabla f(\bar{x}) = 0$ and $\nabla^2 f(\bar{x})$ semipositive definite.) Usually the *p*-regularized subproblem (1.9) is solved by seeking the unique root of a *secular equation* as described by Gould et al. (2010). A theoretical presentation and application of the *p*-regularized subproblems for $p > 2$ was given by Hsia et al. (2017).

1.3   A Short Review of Derivative-free Methods for Unconstrained Optimization

In many optimization problems arising from scientific, engineering, and artificial intelligence applications, the objective function is available only as the output of a black-box or of a simulation oracle that does not provide any derivative information (the gradient or/and the Hessian). Besides, there are many optimization problems, for which their mathematical model is very complicated and consequently, the derivative information of the functions defining the problem is unavailable, impractical to obtain, or unreliable. Such settings need to make use of methods for *derivative-free optimization*, still called *direct search* methods or *zeroth-order* methods, that is, the methods which use only the function values (Conn et al. 2009; Audet and Hare 2017). Derivative-free optimization methods have a long history, dating back to the works of Brooks (1958), Hooke and Jeeves (1961), Spendley et al. (1962), Rastrigin (1963), and Karnopp (1963), and they know a rapid growth, mainly fueled by a growing number of applications that range from science problems (see Abramson et al. 2008; Gray et al. 2004; Zhao et al. 2006) to medical problems (see Marsden et al. 2008) to engineering design and facility location problems (see Abramson and Audet 2006; Audet et al. 2008; Bartholomew-Biggs et al. 2002).

The derivative-free methods can be classified as *direct* and *model-based*. Both of them may be *randomized*. Briefly, direct algorithms determine search directions by computing values of the function f directly, whereas model-based algorithms construct and utilize a surrogate model of the minimizing function to guide the search process. Furthermore, these methods are classified as *local* or *global*. Finally, these algorithms can be *stochastic* or *deterministic*, depending upon whether they require random search steps or not.

Direct search methods include those of a *simplicial* type, where one typically moves away from the worst point (a point at which the function value is the highest), and those of a *directional* type, where one tries to move along a direction defined by the best point.

Simplex methods construct and manipulate a collection of $n + 1$ affinely independent points in $\mathbb{R}^n$ called vertices, which define a simplex and use some rules (reflection, expansion, inner contraction, and shrink) of modifying these vertices of the simplex until the best vertex is determined. The most popular and representative simplex method is that of Nelder and Mead (1965). The version of Nelder and Mead was modified by many researchers. For example, Rykov (1980) proposed a simplex method in which the number of the reflected vertices is modified from iteration to iteration. Tseng (1999) proposed a modified simplex method that keeps a number of the best simplex vertices on a given iteration and uses them to reflect the remaining vertices. Price et al. (2002) embedded the Nelder-Mead algorithm in a different convergent algorithm using positive spanning sets. Nazareth and Tseng (2002) propose a variant that connects the Nelder-Mead algorithm to the golden-section search. Kelley (1999) suggested restarting the Nelder-Mead method when the objective decrease on consecutive iterations is not larger than a multiple of the simplex gradient norm. The convergence of the Nelder-Mead algorithm with appropriate reflection and expansion coefficients in low dimensions was proved by Lagarias et al. (1998).

Each iteration of the *directional direct-search* methods generates a finite set of points around the current point x_k called poll points by taking x_k and adding terms of the form $\alpha_k d$, where α_k is a positive step and d is an element from a finite set of directions D_k, poll direction set. Kolda et al. (2003) called these methods as *generating set-search methods*. The objective function is evaluated at all these poll points, and x_{k+1} is selected as the poll point that produces a sufficient decrease in the objective and possibly the step size α_k is increased. If no poll point provides a sufficient decrease, then x_{k+1} is set to x_k and the step size α_k is decreased. In either case, the set of directions D_k can be modified to get a new set D_{k+1}. These methods differentiate in the way they generate the set of poll directions D_k at every iteration. One of the first directional direct-search methods is the *coordinate search*, in which the poll directions are defined as $D_k = \{\pm e_i : i = 1, \ldots, n\}$, where e_i is the column i of the identity matrix. The first global convergence result for the coordinate search was given by Lewis et al. (2000). A combination of the directional direct-search method and the simplex method is the *pattern-search method* presented by Torczon (1991). In it, for the given simplex defined by $x_k, y_1, \ldots, y_n$ (here x_k is the vertex with the smallest function value), the polling directions are given by $D_k = \{y_i - x_k : i = 1, \ldots n\}$. Torczon shows that if f is continuous on the level set of x_0 and this level set is compact, then there is a subsequence of points $\{x_k\}$ convergent to a stationary point of f. A generalization of the pattern-search method is the *generalized pattern-search* method introduced by Audet (2004). This is characterized by fixing a positive spanning set D and at every iteration, the set D_k is selected from this set as $D_k \subseteq D$. The generalized pattern-search method was analyzed by Dolan et al. (2003), Abramson (2005) and Abramson et al. (2013). A further generalization of the generalized pattern-search method is the *mesh adaptive direct search* (see Audet and Dennis Jr. 2006; Abramson and Audet 2006). A thorough discussion about the mesh adaptive direct search is given by Larson et al. (2019).

The *model-based methods* for derivative-free optimization are based on the predictions of a model that serves as surrogate of the minimizing function. Mainly, these methods use the following approximations: polynomial models, quadratic interpolation models (Winfield 1973; Conn and Toint 1996; Powell 1998, 2002; Fasano et al. 2009; Scheinberg and Toint 2010), underdetermined quadratic interpolation models (Powell 2003, 2004, 2006, 2007, 2008, 2013), regression models (Conn et al. 2008), and radial basis function interpolation models (Buhmann 2000). In this context, polynomial models are the most commonly used. The polynomial methods use the space $P^{d,n}$ of polynomials of n variables of degree d with the basis $\left[1, x_1, \ldots, x_n, x_1^2, \ldots, x_n^2, \ldots, x_1 x_2 \ldots x_{n-1} x_n\right]$. Obviously, the dimension of $P^{2,n}$ is $(n+1)(n+2)/2$. For example, *linear models* can be obtained by using the first $n+1$ components of the above basis. *Quadratic models* with diagonal Hessian can be obtained by using the first $2n+1$ components of the above basis (see Powell 2003). Any polynomial method is defined by the above basis and by a set of coefficients associated to the component of the basis. Given a set of p sample points, the minimizing function is evaluated in these points, and the coefficients of the model are determined by solving a linear algebraic system with p rows and $\dim(P^{d,n})$ columns. The existence, uniqueness, and conditioning of a solution for this system depend on the location of the sample points (see Wendland 2005). Powell (2003, 2004, 2006, 2007, 2008) studied quadratic models constructed from fewer than $(n+1)(n+2)/2$ points. Furthermore, a strategy of using even fewer interpolation points was developed by Powell (2013) and Zhang (2014). Different ways to model the nonlinearity in the frame of derivative-free optimization were developed, for example, by using *radial basis functions* (Buhmann 2000) or *model-based trust-region methods* (see Wild et al. 2008; Marazzi and Nocedal 2002; Maggiar et al. 2018).

The *randomized methods* are very simple to implement and are based on drawing points (possibly uniformly) at random in a domain around the initial point x_0. In the frame of *pure random search*, if $f(x) \leq f(x_0)$, where x is a point randomly generated from the hypersphere of a given radius surrounding the current position, then set $x_0 = x$ and continue the searching. For the convergence of this method, see

Zhigljavsky (1991). Some variants of the random search method subject to the radius of the hypersphere are as follows: fixed step size random search (Rastrigin 1963), optimum step size random search (Schumer and Steiglitz 1968), adaptive step size random search (Schumer and Steiglitz 1968), and optimized relative step size random search (Schrack and Choit 1976). A relatively new random search method is the Nesterov random search (Nesterov and Spokoiny 2017), which is motivated by Gaussian smoothing. The randomization technique was used in the context of directional direct-search or of mesh adaptive direct search. The purpose of the randomized direct-search methods is to improve the direct-search methods, where polling directions are randomly sampled from some distribution at each iteration, hoping to reduce the number of function evaluations (see Audet and Dennis Jr. 2006; Van Dyke and Asaki 2013). Diniz-Ehrhardt et al. (2008) suggested another randomized directional direct-search method that only occasionally employs a descent direction. Bibi et al. (2019) proposed a randomized direct search method in which, at each iteration, the two poll directions are $D_k = \{e_i, -e_i\}$, where e_i is the ith elementary basis vector. At iteration kth, e_i is selected from $\{e_1, \ldots, e_n\}$ with a probability proportional to the Lipschitz constant of the ith partial derivative of the minimizing function. Unlike all these methods, our derivative-free optimization method is a random search method completely different from the known randomized methods.

Excellent reviews and perspectives with an emphasis on highlighting recent developments of the derivative-free optimization methods both for unconstrained and constrained optimization with deterministic, stochastic, or structured objectives were given by Rios and Sahinidis (2013) and by Larson et al. (2019). A review of the derivative-free algorithms, followed by a systematic comparison of 22 related implementations using a test set of 502 problems, was given by Rios and Sahinidis (2013). Their conclusion is not definitive. They emphasize that the ability of all these solvers to obtain good solutions diminishes with the increasing problem size. Besides, attaining the best solutions even for small problems is a challenge for most current derivative-free solvers, and there is no single solver whose performance dominates that of all the others. Dimensionality of the problems and nonsmoothness rapidly increase the complexity of the search and decrease the performances of all solvers. In conclusion, there is a large diversity of derivative-free optimization methods with an impressive development. This is a very active research area, and general textbooks on nonlinear optimization are very quickly outdated in point of this issue.

The *alternatives for derivative-free optimization* methods are classified as the algorithmic differentiation and the numerical differentiation. *Algorithmic differentiation* generates derivatives of mathematical functions that are expressed in computer code (see Griewank 2003; Griewank and Walther 2008). The forward mode of automatic differentiation may be viewed as a differentiation of elementary mathematical operations in each line of the source code by means of the chain rule, while the reverse mode may be seen as traversing the resulting computational graph in reverse order. On the other hand, the *numerical differentiation* estimates the derivative of function f by numerical differentiation and then uses these estimates in a derivative searching method. These numerical differentiation methods were studied both for nonlinear equation and unconstrained optimization (see: Brown and Dennis Jr 1971; Miffin 1975; Dennis and Schnabel 1983; Sun and Yuan 2006; Cartis et al. 2012; Berahas et al. 2019). For the finite-precision functions encountered in scientific applications, finite-difference estimates of derivatives may be sufficient for many purposes.

1.4   Optimality Conditions for Unconstrained Optimization

We are interested in giving conditions under which a solution for the problem (1.1) exists and is qualified as optimal. The purpose is to discuss the main concepts and the fundamental results in unconstrained optimization known as optimality conditions. Both necessary and sufficient conditions for optimality are presented. Plenty of very good books showing these conditions are known: Bertsekas (1999), Nocedal and Wright (2006), Sun and Yuan (2006), Chachuat (2007), Andrei (2017, 2020a), etc. To formulate the optimality conditions, it is necessary to introduce some concepts which characterize an improving direction along which the values of the function f decrease.

Definition 1.1 *(**Descent Direction**): Suppose that $f : \mathbb{R}^n \to \mathbb{R}$ is continuous at x^*. A vector $d \in \mathbb{R}^n$ is a descent direction for f at x^* if there exists $\delta > 0$ so that $f(x^* + \lambda d) < f(x^*)$ for any $\lambda \in (0, \delta)$. The cone of descent directions at x^*, denoted by $C_{dd}(x^*)$ is given by:*

$$C_{dd}(x^*) = \{d : \text{there exists } \delta > 0 \text{ such that } f(x^* + \lambda d) < f(x^*), \text{for any } \lambda \in (0, \delta)\}.$$

Assume that f is a differentiable function. To get an algebraic characterization for a descent direction for f at x^*, let us define the set

$$C_0(x^*) = \{d : \nabla f(x^*)^T d < 0\}.$$

The following result shows that every $d \in C_0(x^*)$ is a descent direction at x^*.

Proposition 1.1 *(**Algebraic Characterization of a Descent Direction**): Suppose that $f : \mathbb{R}^n \to \mathbb{R}$ is differentiable at x^*. If there exists a vector d so that $\nabla f(x^*)^T d < 0$, then d is a descent direction for f at x^*, that is, $C_0(x^*) \subseteq C_{dd}(x^*)$.*

Proof Since f is differentiable at x^*, it follows that

$$f(x^* + \lambda d) = f(x^*) + \lambda \nabla f(x^*)^T d + \lambda \|d\| o(\lambda d),$$

where $\lim_{\lambda \to 0} o(\lambda d) = 0$. Therefore,

$$\frac{f(x^* + \lambda d) - f(x^*)}{\lambda} = \nabla f(x^*)^T d + \|d\| o(\lambda d).$$

Since $\nabla f(x^*)^T d < 0$ and $\lim_{\lambda \to 0} o(\lambda d) = 0$, it follows that there exists a $\delta > 0$ so that $\nabla f(x^*)^T d + \|d\| o(\lambda d) < 0$ for all $\lambda \in (0, \delta)$. ♦

Theorem 1.1 *(First-Order Necessary Conditions for a Local Minimum):* Suppose that $f: \mathbb{R}^n \to \mathbb{R}$ is differentiable at x^*. If x^* is a local minimum, then $\nabla f(x^*) = 0$.

Proof Suppose that $\nabla f(x^*) \neq 0$. Consider $d = - \nabla f(x^*)$, then $\nabla f(x^*)^T d = - \|\nabla f(x^*)\|^2 < 0$. By Proposition 1.1, there exists a $\delta > 0$ so that for any $\lambda \in (0, \delta)$, it follows that $f(x^* + \lambda d) < f(x^*)$. But this is in contradiction with the assumption that x^* is a local minimum for f. ♦

Observe that the above necessary condition represents a system of n algebraic nonlinear equations. All the points x^*, which solve the system $\nabla f(x) = 0$, are called *stationary points*. Clearly, the stationary points need not all be local minima. They could very well be local maxima or even saddle points. In order to characterize a local minimum, we need more restrictive necessary conditions involving the Hessian matrix of the function f.

Theorem 1.2 *(Second-Order Necessary Conditions for a Local Minimum):* Suppose that $f: \mathbb{R}^n \to \mathbb{R}$ is twice differentiable at point x^*. If x^* is a local minimum, then $\nabla f(x^*) = 0$ and $\nabla^2 f(x^*)$ is positive semidefinite.

Proof Consider an arbitrary direction d. Then, using the differentiability of f at x^* we get

$$f(x^* + \lambda d) = f(x^*) + \lambda \nabla f(x^*)^T d + \frac{1}{2} \lambda^2 d^T \nabla^2 f(x^*) d + \lambda^2 \|d\|^2 o(\lambda d),$$

where $\lim_{\lambda \to 0} o(\lambda d) = 0$. Since x^* is a local minimum, $\nabla f(x^*) = 0$. Therefore,

$$\frac{f(x^* + \lambda d) - f(x^*)}{\lambda^2} = \frac{1}{2} d^T \nabla^2 f(x^*) d + \|d\|^2 o(\lambda d).$$

Since x^* is a local minimum, for λ sufficiently small, $f(x^* + \lambda d) \geq f(x^*)$. For $\lambda \to 0$, it follows from the above equality that $d^T \nabla^2 f(x^*) d \geq 0$. Since d is an arbitrary direction, it follows that $\nabla^2 f(x^*)$ is positive semidefinite. ♦

In the above theorems, we have presented the *necessary* conditions for a point x^* to be a local minimum, i.e. these conditions *must be satisfied* at every local minimum solution. However, a point satisfying these necessary conditions need not be a local minimum. In the following theorems, the *sufficient* conditions for a global minimum are given, provided that the objective function is *convex* on $\mathbb{R}^n$. A set $C \subset \mathbb{R}^n$ is a *convex set* if for every points $x, y \in C$ the point $z = \lambda x + (1 - \lambda) y$ is also in the set C for any $\lambda \in [0, 1]$. A function $f: C \to \mathbb{R}$ defined on a convex set $C \subset \mathbb{R}^n$ is a *convex function* if $f(\lambda x + (1 - \lambda) y) \leq \lambda f(x) + (1 - \lambda) f(y)$ for every $x, y \in C$ and every

$\lambda \in (0, 1)$. Moreover, f is said to be *strictly convex* if for every $x, y \in C$ and every $\lambda \in (0, 1)$, $f(\lambda x + (1 - \lambda)y) < \lambda f(x) + (1 - \lambda)f(y)$. The following theorem can be proved. It shows that the convexity is crucial in global nonlinear optimization.

Theorem 1.3 (First-Order Sufficient Conditions for a Strict Local Minimum):
Suppose that $f: \mathbb{R}^n \to \mathbb{R}$ is differentiable at x^* and convex on $\mathbb{R}^n$. If $\nabla f(x^*) = 0$, then x^* is a global minimum of f on $\mathbb{R}^n$.

Proof Since f is convex on $\mathbb{R}^n$ and differentiable at x^*, then from the property of convex functions, it follows that for any $x \in \mathbb{R}^n$ $f(x) \geq f(x^*) + \nabla f(x^*)^T(x - x^*)$. But x^* is a stationary point, that is $f(x) \geq f(x^*)$ for any $x \in \mathbb{R}^n$. ♦

The following theorem gives the second-order sufficient conditions characterizing a local minimum point for those functions which are strictly convex in a neighborhood of the minimum point.

Theorem 1.4 (Second-Order Sufficient Conditions for a Strict Local Minimum).
Suppose that $f: \mathbb{R}^n \to \mathbb{R}$ is twice differentiable at point x^*. If $\nabla f(x^*) = 0$ and $\nabla^2 f(x^*)$ is positive definite, then x^* is a local minimum of f.

Proof Since f is twice differentiable, for any $d \in \mathbb{R}^n$, we can write:

$$f(x^* + d) = f(x^*) + \nabla f(x^*)^T d + \frac{1}{2}d^T \nabla^2 f(x^*)d + \|d\|^2 o(d),$$

where $\lim_{d \to 0} o(d) = 0$. Let λ be the smallest eigenvalue of $\nabla^2 f(x^*)$. Since $\nabla^2 f(x^*)$ is positive definite, it follows that $\lambda > 0$ and $d^T \nabla^2 f(x^*)d \geq \lambda \|d\|^2$. Therefore, since $\nabla f(x^*) = 0$, we can write:

$$f(x^* + d) - f(x^*) \geq \left[\frac{\lambda}{2} + o(d)\right]\|d\|^2.$$

Since $\lim_{d \to 0} o(d) = 0$, then there exists a $\eta > 0$ so that $|o(d)| < \lambda/4$ for any $d \in B(0, \eta)$, where $B(0, \eta)$ is the open ball of radius η centered at 0. Hence

$$f(x^* + d) - f(x^*) \geq \frac{\lambda}{4}\|d\|^2 > 0$$

for any $d \in B(0, \eta) \setminus \{0\}$, i.e. x^* is a strict local minimum of function f. ♦

If we assume f to be twice continuously differentiable, we observe that, since $\nabla^2 f(x^*)$ is positive definite, then $\nabla^2 f(x^*)$ is positive definite in a small neighborhood of x^* and therefore f is strictly convex in a small neighborhood of x^*. Hence, x^* is a strict local minimum, it is the unique global minimum over a small neighborhood of x^*.

Remark 1.1 *As known, a local minimizer of the general function f is a point x^* so that $f(x^*) \leq f(x)$ for all x near x^*. For derivative searching methods, the point x^* is qualified as the local minimum of f if $\nabla f(x^*) = 0$ and $\nabla^2 f(x^*)$ is positive semidefinite.*

These are criteria for the optimality of x^*, implicit imbedded in the derivative searching methods. On the other hand, the derivative-free searching methods which do not use the gradient and the Hessian of function f determine a point x^* for which $f(x^*) \leq f(x)$ for x near x^*, based only on the values of the minimizing function f computed at a sequence of points generated in a specific way. For these methods, nothing can be said about the optimality of x^*. We can only say that $f(x^*) \leq f(x)$ for x near x^*.

1.5   Structure of the Book

The book is structured in five chapters. The first has an introductory character, where the unconstrained optimization problem is defined by presenting the main concepts used for solving this class of optimization problems. The derivative methods for unconstrained optimization are discussed, and then a short review of derivative-free methods is presented. The necessary and sufficient optimality conditions for unconstrained optimization are given, showing that a point x^* is qualified as the local minimum of the function f if $\nabla f(x^*) = 0$ and $\nabla^2 f(x^*) \geq 0$.

Chapter 2 is dedicated to present a two-level random search method for unconstrained optimization. In this method, the minimizing function is supposed to be continuous, lower-bounded, and its minimum value is known. The purpose is to determine the point, where this minimum value is attained. Firstly, a description of the algorithm is given, followed by its presentation on steps. The idea of this algorithm, which is completely different from the other known derivative-free methods for unconstrained optimization, is to catch a deep insight into the landscape of the minimizing function around the initial point by randomly generating some trial points at two levels (Andrei 2020b). At the first level, a number of trial points are randomly generated in a domain around the initial point. The set of all trial points is called a complex. At the second level, in a domain around each trial point from the first level, the algorithm generates another number of trial points, the so-called local trial points. The number of the trial points, the number of the local trial points, and the domains defined by their bounds, in which these trial points are generated, are specified by the user. The minimizing function is evaluated in all these trial points, and by using some specific rules at each iteration, the trial point corresponding to the minimum value of the minimizing function is selected. At each iteration, the algorithm consists of a number of rules for replacing the trial points with local trial points and for generating some new trial points, where the function value is smaller. Thus, using the minimum and the maximum trial points, at which the minimizing function takes its minimum and its maximum value, respectively, the middle point is computed. By using this middle point and a reflection coefficient, the trial points are reflected. Thus, new trial points are obtained at which the minimizing function is likely to be reduced. Furthermore, taking the second minimum trial point, a simple line search is initiated from the current minimum point along the direction determined by the minimum point and the second minimum point. Using this simple line search, the values of the minimizing function might be reduced. Next, the algorithm considers some criteria for reducing (halving) the bounds defining the domains of the trial points or the bounds of the domains of the local trial points, as well as the complex. Details on the algorithm concerning: the choice of the number of trial points and of the number of the local trial points, the choice of the bounds of the domains, where the trial or the local trial points are generated, local versus global character of the searching, the stopping criteria, and the line search are also discussed.

Chapter 3 aims at proving the convergence of the algorithm. For this, the evolution of the maximum distance among the trial points and the evolution of the maximum distance among the trial points and the local points are introduced. It is proved that these distances tend to zero, thus ensuring the convergence of the algorithm. The key element of the algorithm is that it generates a sequence of points $\{x_k\}$ for which the corresponding sequence $\{f(x_k)\}$ is monotonously decreasing and bounded. It follows that the sequence $\{f(x_k)\}$ is convergent to a value $f(x^*)$. Therefore, by continuity, the corresponding sequence $\{x_k\}$ is convergent to a point x^* for which $f(x^*) \leq f(x_0)$. Since there is no access to the derivative information of the minimizing function f, nothing can be said about the optimality of x^*.

In Chap. 4, the numerical performances of the algorithm are reported for solving 140 unconstrained optimization problems, out of which 16 are real applications. The typical evolution of the minimizing function values displays in two forms, showing that the optimization process has two phases: the reduction phase and the stalling one. In the reduction phase, for a relatively small number of iterations, the function values are strongly reduced. In some cases, this reduction presents some short plateaus, where the values of the minimizing function change very slowly. In the stalling phase, for a large number of iterations, the algorithm reduces the function values very slowly. Comparisons with the Nelder-Mead algorithm show that our two-level random search method for the unconstrained optimization algorithm is more efficient and more robust. Finally, the performances of the algorithm for solving a number of 30 large-scale unconstrained optimization problems up to 500 variables are presented. These numerical results show that this approach based on the two-level random search method for unconstrained optimization is able to solve a large diversity of problems with different structures and complexities. However, there are a number of open problems which refer to the following aspects: the selection of the number of trial or of the number of the local trial points, the selection of the bounds of the domains, where the trial points and the local trial points are randomly generated, and a criterion for initiating the line search. Chapter 5 presents the conclusions of this work. Finally, four annexes show the applications and the test functions used in our numerical experiments and illustrate the performances of the algorithm.

References

Abramson, M.A., (2005). Second-order behavior of pattern search. SIAM Journal on Optimization, 16(2), 515-530.

Abramson, M.A., Asaki, T.J., Dennis, J.E. Jr., O'Reilly, K.R., & Pingel, R.L., (2008). Quantitative object reconstruction via Abel-based X-ray tomography and mixed variable optimization. SIAM J. Imaging Sci. 1, 322–342.

Abramson, M.A., & Audet, C., (2006). Convergence of mesh adaptive direct search to second-order stationary points. SIAM Journal on Optimization, 17(2), 606-619.

Abramson, M.A., Frimannslund, L., & Steihaug, T., (2013). A subclass of generating set search with convergence to second-order stationary points. Optimization Methods and Software, 29(5), 900-918.

Andrei, N., (1999). *Programarea Matematică Avansată. Teorie, Metode Computaţionale, Aplicaţii.* [*Advanced Mathematical Programming. Theory, Computational Methods, Applications*] Editura Tehnică [Technical Press], Bucureşti.

Andrei, N., (2009). *Critica Raţiunii Algoritmilor de Optimizare fără Restricţii.* [*Criticism of the Unconstrained Optimization Algorithms Reasoning*]. Editura Academiei Române, Bucureşti.

Andrei, N., (2015). *Critica Raţiunii Algoritmilor de Optimizare cu Restricţii.* [*Criticism of the Constrained Optimization Algorithms Reasoning*]. Editura Academiei Române, Bucureşti.

Andrei, N., (2017). *Continuous Nonlinear Optimization for Engineering Applications in GAMS Technology.* Springer Optimization and Its Applications Series. Vol. 121, New York, NY, USA: Springer Science + Business Media.

Andrei, N., (2020a). *Nonlinear Conjugate Gradient Methods for Unconstrained Optimization.* Springer Optimization and Its Applications Series. Vol. 158, New York, NY, USA: Springer Science + Business Media.

Andrei, N., (2020b). *A simple deep random search method for unconstrained optimization. Preliminary computational results.* Technical Report No.1/2020. AOSR – Academy of Romanian Scientists, Bucharest, Romania, February 29, 2020. (87 pages)

Audet, C., (2004). Convergence results for generalized pattern search algorithms are tight. Optimization and Engineering, 5(2), 101-122.

Audet, C., Béchard, V., & Chaouki, J., (2008). Spent potliner treatment process optimization using a MADS algorithm. Optim. Eng. 9, 143–160.

Audet, C., & Dennis, J.E. Jr., (2006). Mesh adaptive direct search algorithms for constrained optimization. SIAM Journal on Optimization, 17(1), 188-217.

Audet, C., & Hare, W., (2017). *Derivative-Free and Blackbox Optimization.* Springer Series in Operations Research and Financial Engineering. Springer.

Bartholomew-Biggs, M., (2008). *Nonlinear optimization with engineering applications.* New York, NY, USA: Springer Science + Business Media.

Bartholomew-Biggs, M.C., Parkhurst, S.C., & Wilson, S.P. (2002). Using DIRECT to solve an aircraft routing problem. Comput. Optim. Appl. 21, 311–323.

Bazaraa, M.S., Sherali, H.D., & Shetty, C.M., (1993). *Nonlinear Programming Theory and Algorithms.* John Wiley, New York, 2nd edition.

Berahas, A.S., Byrd, R.H., & Nocedal, J. (2019). *Derivative-free optimization of noisy functions via quasi-Newton methods.* SIAM Journal on Optimization, 2019. To appear.

Bertsekas, D.P., (1999). *Nonlinear Programming.* Athena Scientific, Belmont, MA, Second edition.

Bibi, A., Bergou, El H., Sener, O., Ghanem, B., and Richtárik, P., (2019). A stochastic derivative-free optimization method with importance sampling. Technical Report 1902.01272, arXiv. URL https://arxiv.org/abs/1902.01272

Brown, K.M., & Dennis, J.E. Jr. (1971). Derivative free analogues of the Levenberg-Marquardt and Gauss algorithms for nonlinear least squares approximation. Numerische Mathematik, 18(4):289-297.

Broyden, C.G., (1970). The convergence of a class of double-rank minimization algorithms. I. General considerations. Journal of the Institute of Mathematics and Its Applications, 6, 76-90.

Buhmann, M.D., (2000). Radial basis functions. Acta Numerica, 9, 1-38.

Brooks, S.H., (1958). Discussion of random methods for locating surface maxima. Operations Research, 6, 244-251.

Cartis, C., Gould, N.I.M., & Toint, P.L. (2012). On the oracle complexity of first-order and derivative-free algorithms for smooth nonconvex minimization. SIAM Journal on Optimization, 22(1), 66-86.

Cauchy, A., (1847). Méthodes générales pour la resolution des systèmes déquations simultanées. C. R. Acad. Sci. Par. 25(1), 536-538.

Chachuat, B.C., (2007). *Nonlinear and Dynamic Optimization – from Theory to Practice*, IC-31: Winter Semester 2006/2007. École Politechnique Fédérale de Lausanne.

Conn, A.R., & Toint, P.L., (1996). An algorithm using quadratic interpolation for unconstrained derivative free optimization. In G. D. Pillo and F. Giannessi, editors, *Nonlinear Optimization and Applications*, pages 27-47. Springer.

Conn, A.R., Gould, N.I.M., & Toint, Ph.L., (2000). *Trust-Region Methods*. MPS-SIAM Series on Optimization, SIAM, Philadelphia, PA, USA.

Conn, A.R., Scheinberg, K. & Vicente, L.N., (2008). Geometry of sample sets in derivative free optimization: Polynomial regression and underdetermined interpolation. IMA Journal of Numerical Analysis, 28(4), 721-748.

Conn A.R., Scheinberg, K., & Vicente, L.N. (2009). *Introduction to derivative-free optimization*. MPS-SIAM Series on Optimization. SIAM, Philadelphia

Davidon, W.C., (1959). *Variable metric method for minimization*. Research and Development Report ANL-5990. Argonne National Laboratories.

Dembo, R.S., Eisenstat, S.C., & Steihaug, T., (1982). Inexact Newton methods. SIAM Journal on Numerical Analysis 19, 400-408.

Dembo, R.S., & Steihaug, T., (1983). Truncated Newton algorithms for large-scale unconstrained optimization. Mathematical Programming, 26, 190-212.

Dennis, J.E., & Schnabel, R.B., (1983). *Numerical Methods for Unconstrained Optimization and Nonlinear Equations*. Prentice-Hall, Englewoods Cliffs, New Jersey. [Reprinted as Classics in Applied Mathematics 16, SIAM, Philadelphia, USA, 1996.]

Deuflhard, P., (1990). Global inexact Newton methods for very large scale nonlinear problems. In Proceedings of the Cooper Mountain Conference on Iterative Methods. Cooper Mountain, Colorado, April 1-5.

Diniz-Ehrhardt, M.A., Martínez, J.M., & Raydan, M., (2008). A derivative-free nonmonotone linesearch technique for unconstrained optimization. Journal of Computational and Applied Mathematics, 219(2), 383-397.

Dolan, E.D., Lewis, R.M., & Torczon, V., (2003). On the local convergence of pattern search. SIAM Journal on Optimization, 14(2), 567-583.

Fasano, G., Morales, J.L., & Nocedal, J., (2009). On the geometry phase in model-based algorithms for derivative-free optimization. Optimization Methods and Software, 24(1), 145-154.

Fletcher, R., (1970). A new approach to variable metric algorithms. The Computer Journal, 13, 317-322.

Gill, P.E., Murray, W., & Wright, M.H., (1981). *Practical Optimization*. Academic Press, New York.

Goldfarb, D., (1970). A family of variable metric method derived by variation mean. Mathematics of Computation, 23, 23-26.

Gould, N.I.M., Robinson, D.P., & Sue Thorne, H., (2010). On solving trust-region and other regularized subproblems in optimization. Mathematical Programming Computation, 2(1), 21-57.

Gray, G., Kolda, T., Sale, K., & Young, M., (2004). Optimizing an empirical scoring function for transmembrane protein structure determination. INFORMS J. Comput. 16, 406–418.

Griewank, A., (1981). *The modification of Newton's method for unconstrained optimization by bounding cubic terms*. Technical Report NA/12, Department of Applied Mathematics and Theorerical Physics, University of Cambridge.

Griewank, A. (2003). A mathematical view of automatic differentiation. Acta Numerica, 12:321-398.

Griewank, A., & Walther, A. (2008). *Evaluating Derivatives: Principles and Techniques of Algorithmic Differentiation*. Second edition, SIAM – Society for Industrial and Applied Mathematics, Philadelphia, USA.

Hooke, R. & Jeeves, T.A. (1961). Direct search solution of numerical and statistical problems. Journal of the ACM. 8(2): 212–229.

Hsia, Y., Sheu, R.L., & Yuan, Y.X. (2017). Theory and application of p-regularized subproblems for $p>2$. Optimization Methods & Software, 32(5), 1059–1077.

Karnopp, D.C., (1963). Random search techniques for optimization problems. Automatica, 1, 111-121.

Kelley, C.T., (1999). Detection and remediation of stagnation in the Nelder-Mead algorithm using a sufficient decrease condition. SIAM Journal on Optimization, 10(1), 43-55.

Kolda, T.G., Lewis, R.M., & Torczon, V., (2003). Optimization by direct search: New perspectives on some classical and modern methods. SIAM Review, 45(3):385-482.

Lagarias, J.C., Reeds, J.A., Wright, M.H., & Wright, P.E., (1998). Convergence properties of the Nelder-Mead simplex algorithm in low dimensions. SIAM Journal on Optimization, 9, 112-147.

Larson, J., Menickelly, M., & Wild, S.M., (2019). *Derivative-free optimization methods*. Mathematics and Computer Science Division, Argonne National Laboratory, Lemont, IL 60439, USA, April 4, 2019.

Lewis, R.M., Torczon, V., & Trosset, M.W., (2000). Direct search methods: Then and now. Journal of Computational and Applied Mathematics, 124(1-2), 191-207.

Luenberger, D.G., (1984). *Introduction to linear and nonlinear programming*. Addison-Wesley Publishing Company, Reading, Second edition, 1984.

Luenberger, D. G., & Ye, Y., (2016). *Linear and nonlinear programming*. International Series in Operations Research & Management Science 228 (Fourth edition). New York, Springer.

Maggiar, A., Wächter, A., Dolinskaya, I.S., & Staum, J., (2018). A derivative-free trust-region algorithm for the optimization of functions smoothed via Gaussian convolution using adaptive multiple importance sampling. SIAM Journal on Optimization, 28(2):1478-1507.

Marazzi, M., & Nocedal, J., (2002). Wedge trust region methods for derivative free optimization. Mathematical Programming, 91(2):289-305.

Marsden, A.L., Feinstein, J.A., & Taylor, C.A., (2008). A computational framework for derivative-free optimization of cardiovascular geometries. Comput. Methods Appl. Mech. Eng. 197, 1890–1905.

Miffin, R. (1975). A superlinearly convergent algorithm for minimization without evaluating derivatives. Mathematical Programming, 9(1), 100-117.

Nazareth, L., & Tseng, P., (2002). Gilding the lily: A variant of the Nelder-Mead algorithm based on golden-section search. Computational Optimization and Applications, 22(1), 133-144.

Nelder, J.A., & Mead, R., (1965). A simplex method for function minimization. The Computer Journal, 7(4), 308-313.

Nesterov, Y., & Spokoiny, V., (2017). Random gradient-free minimization of convex functions. Foundations of Computational Mathematics, 17(2), 527-566.

Nocedal, J., (1980). Updating quasi-Newton matrices with limited storage. Mathematics of Computation 35, 773-782.

Nocedal, J., & Wright, S.J., (2006). *Numerical optimization*. Springer Series in Operations Research. Springer Science+Business Media, New York, Second edition, 2006.

Powell, M.J.D., (1970). A new algorithm for unconstrained optimization. In J.B. Rosen, O.L. Mangasarian and K. Ritter (Eds.) *Nonlinear Programming*. Academic Press, New York, 31-66.

Powell, M.J.D., (1998). Direct search algorithms for optimization calculations. Acta Numerica, 7, 287-336.

Powell, M.J.D., (2002). UOBYQA: unconstrained optimization by quadratic approximation. Mathematical Programming, 92, 555-582.

Powell, M.J.D., (2003). On trust region methods for unconstrained minimization without derivatives. Mathematical Programming, 97, 605-623.

Powell, M.J.D., (2004). Least Frobenius norm updating of quadratic models that satisfy interpolation conditions. Mathematical Programming, 100(1), 183-215.

Powell, M.J.D., (2006). The NEWUOA software for unconstrained optimization without derivatives. In G. D. Pillo and M. Roma, editors, *Large-Scale Nonlinear Optimization*, volume 83 of Nonconvex Optimization and its Applications, pages 255-297. Springer.

Powell, M.J.D., (2007). *A view of algorithms for optimization without derivatives*. Technical Report DAMTP 2007/NA03, University of Cambridge.

Powell, M.J.D., (2008). Developments of NEWUOA for minimization without derivatives. IMA Journal of Numerical Analysis, 28(4), 649-664.

Powell, M.J.D., (2013). Beyond symmetric Broyden for updating quadratic models in minimization without derivatives. Mathematical Programming, 138(1-2), 475-500.

Price, C.J., Coope, I.D., & Byatt, D., (2002). A convergent variant of the Nelder-Mead algorithm. Journal of Optimization Theory and Applications, 113(1), 5-19.

Rastrigin, L.A., (1963). The convergence of the random search method in the extremal control of a many parameter system. Automation and Remote Control. 24 (10), 1337–1342.

Rykov, A.S., (1980). Simplex direct search algorithms. Automation and Remote Control, 41(6), 784-793.

Rios, L.M., & Sahinidis, N.V., (2013). Derivative-free optimization: a review of algorithms and comparison of software implementations. Journal of Global Optimization, 56, 1247-1293.

Scheinberg, K., & Toint, P.L., (2010). Self-correcting geometry in model-based algorithms for derivative-free unconstrained optimization. SIAM Journal on Optimization, 20(6), 3512-3532.

Schumer, M.A., & Steiglitz, K., (1968). Adaptive step size random search. IEEE Transactions on Automatic Control. 13(3), 270–276.

Schrack, G. & Choit, M., (1976). Optimized relative step size random searches. Mathematical Programming. 10(1), 230–244.

Shanno, D.F., (1970). Conditioning of quasi-Newton methods for function minimization. Mathematics of Computation, 24, 647-656.

Spendley, W., Hext, G.R., & Himsworth, F.R., (1962). Sequential application for simplex designs in optimization and evolutionary operation. Technometrics 4, 441–461.

Sun, W., & Yuan, Y.X., (2006). *Optimization Theory and Methods. Nonlinear Programming*. Springer Science + Business Media, New York, 2006.

Tseng, P., (1999). Fortified-descent simplicial search method: A general approach. SIAM Journal on Optimization, 10(1), 269-288.

Torczon, T., (1991). On the convergence of the multidirectional search algorithm. SIAM Journal on Optimization, 1(1), 123-145.

Van Dyke, B., & Asaki, T.J., (2013). Using QR decomposition to obtain a new instance of mesh adaptive direct search with uniformly distributed polling directions. Journal of Optimization Theory and Applications, 159(3), 805-821.

Wendland, H., (2005). Scattered Data Approximation. Cambridge Monographs on Applied and Computational Mathematics. Cambridge University Press.

Wild, S.M., Regis, R.G., & Shoemaker, C.A., (2008). ORBIT: Optimization by radial basis function interpolation in trust-regions. SIAM Journal on Scientific Computing, 30(6), 3197-3219.

Winfield, D.H., (1973). Function minimization by interpolation in a data table. Journal of the Institute of Mathematics and its Applications, 12, 339-347.

Zhang, Z., (2014). Sobolev seminorm of quadratic functions with applications to derivative-free optimization. Mathematical Programming, 146(1-2), 77-96.

Zhao, Z., Meza, J.C., & Van Hove, M., (2006). Using pattern search methods for surface structure determination of nanomaterials. J. Phys. Condens. Matter 18, 8693–8706.

Zhigljavsky, A.A., (1991). *Theory of Global Random Search*. Springer Netherlands.

Chapter 2
A Derivative-Free Two-Level Random Search Method for Unconstrained Optimization

Abstract The purpose of this chapter is to present a two-level random search method for unconstrained optimization and the corresponding algorithm. The idea of the algorithm is to randomly generate a number of trial points in some domains at two levels. At the first level, a number of trial points are generated around the initial point, where the minimizing function is evaluated. At the second level, another number of local trial points are generated around each trial point, where the minimizing function is evaluated again. The algorithm consists of a number of rules for replacing the trial points with local trial points and for generating some new trial points to get a point where the function value is smaller. Some details of the algorithm are developed and discussed, concerning the number of trial points, the number of local trial points, the bounds of the domains where these trial points are generated, the reduction of the bounds of these domains, the reduction of the trial points, middle points, the local character of searching, the finding of the minimum points, and the line search used for accelerating the algorithm. The numerical examples illustrate the characteristics and the performances of the algorithm. At the same time, some open problems are identified.

Keywords Trial points · Local trial points · Complex · Random points · Bounds · Reflection point · Middle point · Line search

Let us consider the problem: ·

$$\min f(x), \tag{2.1}$$

where $f : \mathbb{R}^n \to \mathbb{R}$, is bounded below on $\mathbb{R}^n$ and n is relatively small, let us say $n \leq 500$. Many algorithms have been proposed to solve this problem, but in this book, we are interested in dealing with the case in which the derivatives of this function are unavailable, impractical to obtain, or unreliable. As we have already seen, there are a lot of randomized methods based on different strategies using pure random search, randomized directional direct-search, randomized mesh adaptive direct-search, or some variants of them. In the following, we present a new derivative-free random search method at two levels, which is completely different from the known randomized methods for unconstrained optimization. Roughly speaking, our method is close to pure random search, but extended at two levels.

Suppose that all the variables are in a domain $D \subset \mathbb{R}^n$ defined by some bounds on variables. Some of these bounds are imposed by the physical constraints which define the problem. Others are artificially introduced for being used in our algorithm. Anyway, we consider that all the variables are bounded by some known bounds, the same for all of them. Suppose that the minimum value of the function f is known as f^{opt}. Assume that the function f is continuous and bounded below on $\mathbb{R}^n$. This chapter presents a new derivative-free algorithm based on randomly generating some trial points in the domain D.

2.1   Description of the Algorithm

The idea of the algorithm is as follows: Set the current iteration $iter = 1$ and assign a value to the maximum number of iterations admitted by the algorithm as max$iter$. Suppose that the searching of the minimum of the function f starts from the initial point $x_0 \in \mathbb{R}^n$. Consider an initial domain $D \subset \mathbb{R}^n$ defined by its bounds (lower and upper) to each of the n variables. These limits may be the same for all the variables, let us say $lobnd$ and $upbnd$. In other words, if on components we have $x_i = \left[x_i^1, \ldots, x_i^n \right]$, then assume that $lobnd \leq x_i^l \leq upbnd$ for all $l = 1, \ldots, n$. Of course, the bounds $lobnd$ and $upbnd$ can be different for different components of the variables x_i, but in the following, we assume that they are the same for each component x_i^l, $l = 1, \ldots, n$. We emphasize that these bounds $lobnd$ and $upbnd$ maybe the bounds defining the optimization problem and imposed by the engineering constructive specifications of the minimizing problem.

Around the initial point x_0, we randomly generate over D a number N of trial points, $x_1, \ldots, x_N$, where $x_j \in \mathbb{R}^n$, for $j = 1, \ldots, N$. If $N = n + 1$, then these points define a simplex in R^n, that is, the convex hull of its $n + 1$ points $x_1, \ldots, x_{n+1}$. Otherwise, if $N \neq n + 1$, we say that we have a geometric structure, which we call a *complex* in $\mathbb{R}^n$. Obviously, N may be larger or smaller than the number of variables n, or even equal to n. Evaluate the minimizing function f in these points, thus obtaining $f(x_1), \ldots, f(x_N)$.

Now, for every trial point $x_j, j = 1, \ldots, N$, let us define the local domains $D_j \subset \mathbb{R}^n$ by specifying the bounds, let us say $lobndc$ and $upbndc$. Of course, these bounds may be different for every trial point, but in our development, we consider them equal for every trial point. Using these bounds $lobndc$ and $upbndc$ around each trial point x_j, we randomly generate M local trail points $x_j^1, \ldots x_j^M$, where $x_j^i \in \mathbb{R}^n$, for $i = 1, \ldots, M$. In other words, around the trial point x_j, the local trial points $x_j^1, \ldots x_j^M$ are bounded by the limits $lobndc$ and $upbndc$. Usually, the local domains D_j are smaller than the domain D, but this is not compulsory. Now, evaluate the function f in these local trial points, thus obtaining $f\left(x_j^1\right), \ldots, f\left(x_j^M\right)$ for any $j = 1, \ldots, N$.

With this, determine the local point x_j^k corresponding to the minimum value from the set $\left\{ f\left(x_j^1\right), \ldots, f\left(x_j^M\right) \right\}$, that is, $f\left(x_j^k\right) = \min \left\{ f\left(x_j^1\right), \ldots, f\left(x_j^M\right) \right\}$. The following decision is now taken: If $f\left(x_j^k\right) < f\left(x_j\right)$, then replace the trial point x_j with the local trial point x_j^k and the value $f(x_j)$ with $f\left(x_j^k\right)$.

The above procedure is repeated for every trial point $x_j, j = 1, \ldots, N$. At the end of this cycle, if the case, the trial points $x_j, j = 1, \ldots, N$, are replaced by the local trial points $x_j^k, j = 1, \ldots, N$, for which the function value $f\left(x_j^k\right)$ might be smaller than $f(x_j)$. Let us denote these new trial points by $y_1, \ldots, y_N$, that is, $y_j = x_j^k, j = 1, \ldots, N$, and the function values in these points by $f(y_1), \ldots, f(y_N)$, that is, $f\left(y_j\right) = f\left(x_j^k\right)$, $j = 1, \ldots, N$.

Now, determine the point y_k corresponding to the minimum values of the minimizing function, that is, $f(y_k) = \min \{f(y_1), \ldots, f(y_N)\}$.

For $j = 1, \ldots, N, j \neq k$, where k is the index corresponding to the minimum value $f(y_k)$, compute the middle point $z_j = (y_k + y_j)/2$ and evaluate the minimizing function in this point, thus obtaining $f(z_j)$. For $j = 1, \ldots, N, j \neq k$, if $f(z_j) < f(y_j)$, then replace the point y_j with z_j and the value $f(y_j)$ with $f(z_j)$, otherwise do not replace them. Set $z_k = y_k$ and $f(z_k) = f(y_k)$. At this step of the algorithm, another set of new trial points denoted as $z_1, \ldots, z_N$ is obtained, for which the function values $f(z_1), \ldots, f(z_N)$ might be smaller than the previous values of the minimizing function f.

Now, determine $f^{\max} = \max \{f(z_1), \ldots, f(z_N)\}$, corresponding to the trial point $z^{\max}$ from the set of points $\{z_1, \ldots, z_N\}$. Similarly, determine $f^{\min} = \min \{f(z_1), \ldots, f(z_N)\}$, corresponding to the trial point $z^{\min}$ from the set of points $\{z_1, \ldots, z_N\}$. Compute the middle point: $z^m = (z^{\max} + z^{\min})/2$ and the function value in this point $f(z^m)$.

In the following, for $j = 1, \ldots, N$, where the index j is different from the indices corresponding to the minimum and maximum points $z^{\min}$ and $z^{\max}$ respectively, if $f(z^m) < f(z_j)$, then compute the reflected point $z^r = \alpha z^m - z_j$, and if $f(z^r) < f(z_j)$, then replace z_j by the reflected point z^r, that is, set $z_j = z^r$ and $f(z_j) = f(z^r)$. Notice that α is a positive coefficient known as the reflection coefficient, often selected as $\alpha \in [1, 2]$, usually close to 2.

At the end of this cycle, some of the trial points $z_j, j = 1, \ldots, N$, have been replaced by their reflected points. Therefore, we have a new set of trial points $\{z_1, \ldots, z_N\}$, for which the function values $\{f(z_1), \ldots, f(z_N)\}$ might be smaller than the previous values.

Intensive numerical experiments have shown that the values of the minimizing function are significantly reduced in the first few iterations, after which the algorithm is stalling, that is, the function values are very slowly reduced along the iterations. In order to accelerate the algorithm, the following step is implemented. If the difference between the minimum values of the minimizing function at two successive iterations is smaller than a positive, small enough threshold δ, then a simple line search with a positive step β or with a negative step β is initiated from the current minimum point along the direction determined by the minimum point and the second minimum point. Thus, if the function value is reduced, then the minimum point is replaced by the new trial point. Again, at the end of this step, it is possible for the minimum trial point z^{min} to be replaced by another point, for which the function value is smaller. Therefore, another set of trial points $\{z_1, \ldots, z_N\}$ is obtained, for which the function values $\{f(z_1), \ldots, f(z_N)\}$ might be smaller than the previous values.

Now, determine $f^{max} = \max \{f(z_1), \ldots, f(z_N)\}$, corresponding to the trial point z^{max} from the set of trial points $\{z_1, \ldots, z_N\}$. Similarly, determine $f^{min} = \min \{f(z_1), \ldots, f(z_N)\}$, corresponding to the trial point z^{min} from the set of trial points $\{z_1, \ldots, z_N\}$.

From this set of trial points $\{z_1, \ldots, z_N\}$, at the iteration $iter$, determine the point z_k corresponding to the minimum values of the minimizing function, that is, $f(z_k) \equiv f_{iter} = \min \{f(z_1), \ldots, f(z_N)\}$.

At the iteration $iter$, compute the difference: $d_{iter} = \left| f(z_k) - f^{opt} \right| \equiv \left| f_{iter} - f^{opt} \right|$. With these, the following decisions are implemented:

- If $\left| f^{max} \right| > B$, where B is a positive constant sufficiently large, then reduce the bounds $lobnd$ and $upbnd$, as, for example, $lobnd/2$ and $upbnd/2$, respectively. Set $iter = iter + 1$ and go to generate around the initial point x_0 a new set of trial points by using the reduced bounds $lobnd/2$ and $upbnd/2$. At the end of this decision, the bounds $lobnd$ and $upbnd$ are likely to be reduced.

- If $d_{iter} \neq d_{iter - 1}$, then, if $d_{iter} > \varepsilon$, set $iter = iter + 1$ and $d_{iter - 1} = d_{iter}$. If $iter > \max iter$, then go to the final step, else go to the next step. If $d_{iter} = d_{iter - 1}$, then reduce the bounds of the local domain, that is, set $lobndc = lobndc/2$ and $upbndc = upbndc/2$. If $d_{iter} > \varepsilon$, then set $iter = iter + 1$, and if $iter > \max iter$, then go to the final step, else go to the next step. At the end of this decision, the lower and upper bounds of the local domains $lobndc$ and $upbndc$ might be reduced.

- Here we have the next step, where the following decision is taken: If $\left| f_{iter} - f_{iter - 1} \right| < t$, where t is a prespecified small threshold, then reduce the complex, that is, for $j = 1, \ldots, N$, compute $z_j = z^{min} + (z_j - z^{min})/2$.

At this iteration, set $f_{iter - 1} = f_{iter}$, and for $j = 1, \ldots, N$, set $x_j = z_j$ and go to the next iteration by generating M local trail points $x_j^1, \ldots x_j^M$ around each new trial points $x_j, j = 1, \ldots, N$.

The final step: Improve the minimum point obtained as above by randomly selecting a number Q of points in a domain C defined by the bounds $lobndcc$ and $upbndcc$, smaller than $lobndc$ and $upbndc$, respectively. Evaluate the minimizing function in these Q points and select the minimum value from them, which corresponds to the solution of the problem.

More exactly, on steps, the algorithm may be described as follows.

2.2   DEEPS Algorithm

With these developments, the algorithm may be presented on steps as follows:

1.	Initialization. Select: an initial point x_0, the number N of a set of trial points, and the number M of the set of local trial points around each trial point. Select $\varepsilon > 0$ small enough. Select the bounds *lobnd* and *upbnd* which determine the set of trial points, the bounds *lobndc* and *upbndc* which determine the set of local trail points, the bounds *lobndcc* and *upbndcc* which determine the final trail points around the minimum, as well as the threshold t small enough. Select B a positive constant sufficiently large. Select a value for the coefficient α used in the reflection of points and a value for the coefficient β used in a line search along the direction determined by the minimum trial point and by the second minimum trial point. Select a value for the parameter δ used in initiating the above line search. Set *iter* = 1 and max*iter* as the maximum number of iterations admitted by the algorithm.
2.	Around x_0, randomly generate N trial points $x_1, \ldots, x_N$ in a domain D defined by the lower bound *lobnd* and by the upper bound *upbnd* on variables.
3.	Evaluate the minimizing function f in the trial points $f(x_1), \ldots, f(x_N)$.
4.	For $j = 1, \ldots, N$, around x_j do: (a) Generate M points $x_j^1, \ldots x_j^M$ in the domains D_j defined by the bounds *lobndc* and *upbndc*, smaller than the bounds of the domain D. Evaluate the function in these points, thus obtaining the values: $f\left(x_j^1\right), \ldots, f\left(x_j^M\right)$. (b) Determine the point x_j^k corresponding to the minimum values $$f\left(x_j^k\right) = \min\left\{ f\left(x_j^1\right), \ldots, f\left(x_j^M\right)\right\}.$$ (c) If $f\left(x_j^k\right) < f\left(x_j\right)$, then replace the trial point x_j by x_j^k and the value $f(x_j)$ by $f\left(x_j^k\right)$. End for *(By these replacements, at the end of this cycle, another set of new trial points denoted as $y_1, \ldots, y_N$ is obtained, for which the function values are $f(y_1), \ldots, f(y_N)$.)*
5.	Determine the point y_k corresponding to the minimum values of the minimizing function: $f(y_k) = \min\{f(y_1), \ldots, f(y_N)\}$.
6.	For $j = 1, \ldots, N, j \neq k$, do: (a) Compute the point $z_j = (y_k + y_j)/2$ (b) Evaluate the minimizing function in this point, thus obtaining $f(z_j)$ (c) If $f(z_j) < f(y_j)$, then replace the point y_j by z_j and the value $f(y_j)$ by $f(z_j)$ End for
7.	Set $z_k = y_k$ and $f(z_k) = f(y_k)$. *(At the end of steps 6 and 7, another set of new trial points denoted as $z_1, \ldots, z_N$ is obtained, for which the function values are $f(z_1), \ldots, f(z_N)$.)*
8.	Determine $f^{max} = \max\{f(z_1), \ldots, f(z_N)\}$ and the corresponding trial point z^{max}. Determine $f^{min} = \min\{f(z_1), \ldots, f(z_N)\}$ and the corresponding trial point z^{min}.
9.	Compute the middle point $z^m = (z^{max} + z^{min})/2$ and the function value in this point $f(z^m)$.

10.	For $j = 1, \ldots, N$, where the index j is different from the indices corresponding to the minimum and maximum points z^{min} and z^{max} respectively, do:
	If $f(z^m) < f(z_j)$, then
	  Compute the reflected point $z^r = \alpha z^m - z_j$ and $f(z^r)$.
	  If $f(z^r) < f(z_j)$, then
	    Set $z_j = z^r$ and $f(z_j) = f(z^r)$
	  End if
	End if
	End for
	(At the end of this cycle, another set of trial points $z_1, \ldots, z_N$ is obtained, for which the function values are $f(z_1), \ldots, f(z_N)$.)

11.	At iteration *iter* determine: $f_{iter}^{max} = \max\{ f(z_1), \ldots, f(z_N)\}$ and the corresponding trial point z^{max}. $f_{iter}^{min} = \min\{ f(z_1), \ldots, f(z_N)\}$ and the corresponding trial point z^{min}.

(continued)

12.	If $\left\| f_{iter}^{\min} - f_{iter-1}^{\min} \right\| \le \delta$, then do:
	   (a) In the current set of trial points, determine the second minimum trial
	        point z^{mins}
	   (b) Compute the point $z^s = z^{\min} + \beta(z^{\min} - z^{\mathrm{mins}})$
	   (c) Compute $f^s = f(z^s)$
	   If $f^s < f_{iter}^{\min}$, then
	        Set $z^{\min} = z^s$, $f_{iter}^{\min} = f^s$ and go to step 13
	   Else
	       (d) Compute the point $z^s = z^{\min} - \beta(z^{\min} - z^{\mathrm{mins}})$
	       (e) Compute $f^s = f(z^s)$
	       If $f^s < f_{iter}^{\min}$, then
	           Set $z^{\min} = z^s$, $f_{iter}^{\min} = f^s$ and go to step 13
	       Else
	           Go to step 13
	       End if
	   End if
	End if

13.	If $upbnd < 1$ and $lobnd > -1$, then go to step 15.

14.	If $\left\| f_{iter}^{\max} \right\| > B$, then:
	   (a) Reduce the bounds as $lobnd/2$ and $upbnd/2$
	   (b) Set $iter = iter + 1$. If $iter > \max\ iter$, then go to step 19, else go to step 2
	End if

15.	At iteration $iter$, compute the difference: $\left\| f_{iter}^{\min} - f^{opt} \right\|_{iter}$

16.	If $\left\| f_{iter}^{\min} - f^{opt} \right\|_{iter} \ne \left\| f_{iter-1}^{\min} - f^{opt} \right\|_{iter-1}$, then
	   If $\left\| f_{iter}^{\min} - f^{opt} \right\|_{iter} > \varepsilon$, then
	        Set $iter = iter + 1$ and $\left\| f_{iter-1}^{\min} - f^{opt} \right\|_{iter-1} = \left\| f_{iter}^{\min} - f^{opt} \right\|_{iter}$
	        If $iter > \max\ iter$, then go to step 19, else go to step 17.
	   Else
	        Go to step 19
	   End if
	Else
	    Set $lobndc = lobndc/2$ and $upbndc = upbndc/2$
	    If $\left\| f_{iter}^{\min} - f^{opt} \right\|_{iter} > \varepsilon$, then
	        Set $iter = iter + 1$
	        If $iter > \max\ iter$, then go to step 19, else go to step 17
	    Else
	        Go to step 19
	    End if
	End if

17.	If $\left\| f_{iter}^{\min} - f_{iter-1}^{\min} \right\| < t$, then reduce the complex:
	        For $j = 1, \ldots, N$
	           compute $z_j = z^{\min} + (z_j - z^{\min})/2$
	        End for
	End if

18.	Set $f_{iter-1}^{\min} = f_{iter}^{\min}$ For $j = 1, \ldots, N$, set $x_j = z_j$ and go to step 4
19.	Improve the search. Around the minimum point z_k obtained so far, randomly select a number Q of points in the domain C defined by the bounds $lobndcc$ and $upbndcc$, smaller than $lobndc$ and $upbndc$, respectively. Evaluate the minimizing function in these Q points; and from them, select the minimum as the solution of the problem. ◆

2.3 Details on the Algorithm

The algorithm is heuristic. It is based only on comparing the function values at a set of points randomly generated in a domain $D \subset \mathbb{R}^n$, defined by the bounds *lobnd* and *upbnd*.

In this form, the algorithm depends on the number N of the trial points randomly generated around the initial point x_0, on the number M of the local trial points generated around the trial points, as well as on the bounds defining the domain D and the local domains $D_j, j = 1, \ldots, N$.

To ensure the convergence, the algorithm implements three actions in steps 14, 16, and 17.

1. At step 14, if the maximum absolute value of the minimizing function f computed in the current trial points is greater than a prespecified value B sufficiently large, let us say $B = 10^5$, then the bounds *lobnd* and *upbnd* are halved. Therefore, the algorithm generates a sequence of points in bounded domains.
2. Based on the differences between the minimum values of the minimizing function and the known optimum value at two successive iterations, step 16 implements a procedure for halving the bounds *lobndc* and *upbndc* of the local domains, thus shrinking the local search domains.
3. If the difference between the minimum values of the minimization function at two successive iterations is smaller than a threshold t, then the algorithm reduces the complex in step 17.

Let us define BR as the number of reductions of the bounds *lobnd* and *upbnd* (see step 14), BCR as the number of reductions of the local bounds *lobndc* and *upbndc* (see step 16), and CR as the number of complex reductions (see step 17).

The algorithm keeps a balance between the threshold δ for initiating the line searches in step 12 and the threshold t for reducing the complex in step 17. In general, the values for δ are selected as being smaller than those of t. Usually, $\delta = 10^{-4}$ and $t = 10^{-2}$, but some other values are acceptable.

Numerical experiments have shown that the algorithm has two phases: the *reduction phase* and the *stalling* one. In the reduction phase, for some problems, the function values are significantly reduced for the first few iterations. After that, the algorithm gets into the stalling phase, where the function values are reduced very slowly. Both thresholds δ and t are introduced to accelerate the algorithm, that is, to diminish the stalling. Depending on the value of the threshold δ, a line search is initiated from the minimum point along the direction determined by the minimum point and the second minimum point, corresponding to the current set of trial points. Similarly, depending on the value of the threshold t, the algorithm reduces the complex.

In order to accelerate the convergence of the algorithm at every iteration, a number of procedures based on computing the middle point between the minimum point and the current point of the complex and the reflection of points subject to the middle point between the minimum and the maximum points are used. In the steps

9 and 10, the middle point is computed by taking into consideration the points corresponding to the minimum and maximum values of the minimizing function. Subject to this middle point, the reflections of the trial points are executed according to the function values in the reflected points. The reflection coefficient $\alpha \in [1, 2]$ is assigned to a positive value, close to 2. Having in view the formula for computing the reflected point, it is quite possible that this reflected point might not belong to the domain D. In order to give more room to the algorithm, the reflected point is not constrained to be in the domain D. To accelerate the algorithm, at step 12, a line search is initiated from the minimum point along a direction determined by the minimum point and by the second minimum point in the current set of trial points. In other words, if the difference between the minimum values of the minimization function at two successive iterations is smaller than a threshold δ, then a line search is initiated.

Step 19 is included in DEEPS for refining the solution. In this step, around the minimum point obtained so far, a number Q of points are randomly generated in a domain C defined by the bounds *lobndcc* and *upbndcc*, which are small enough. In our numerical experiments, we have considered *lobndcc* $=$ $-$ 0.0001 and *upbndcc* $=$ $+$ 0.0001. Evaluate the minimizing function at all the points from this set and select the one corresponding to the minimum value of the function f.

2.3.1   The Choice of N and M

For a given problem, there is no known optimum value for N and M. These depend on the complexity of the function to be minimized and on the size of the domains D and $D_j, j = 1, \ldots, N$. There should be a balance between the values of N and M. Our intensive numerical experiments have shown that good results are obtained for N smaller that M, but this is not a rule. On the other hand, the procedure for choosing N is dependent on the size of the domain D. Larger domains D recommend larger N. Anyway, the appropriate choice of N and M is a matter of experience, and for getting good results, it is better to select N and M as being smaller, assuming that thoroughness of search is more important than the speed of convergence.

Let us consider an example which illustrates the selection of N and M by minimizing the function:

$$f(x) = \sum_{i=1}^{n} (x_i - 1)^2 + \left(\sum_{i=1}^{n} ix_i - \frac{n(n+1)}{2} \right)^2 + \left(\sum_{i=1}^{n} ix_i - \frac{n(n+1)}{2} \right)^4, \quad (2.2)$$

with the initial point: $x_0^i = 1 - \frac{i}{n}$, $i = 1, \ldots, n$, known as the problem VARDIM (Bongartz et al. 1995). Consider $n = 10$, *lobnd* $= 0$, *upbnd* $= 1$, *lobndc* $= -0.1$, *upbndc* $= +0.1$, and $\varepsilon = 10^{-5}$. Table 2.1 shows the performances of DEEPS for different values of N and M, where *iter* is the number of iteration, *nfunc* is the

Table 2.1 Performances of DEEPS for minimizing the function VARDIM
$n = 10, f(x_0) = 0.219855E + 07$

N	M	iter	nfunc	cpu	BR	BCR	CR	$f(x^*)$
3	3	407	5139	0.01	1	11	378	0.241809E−07
5	2	169	2931	0.00	1	9	150	0.597735E−07
5	3	120	2656	0.00	1	8	102	0.371821E−07
5	5	85	2731	0.00	1	8	69	0.887047E−07
2	5	134	1612	0.00	1	8	118	0.591003E−07
10	5	54	3566	0.01	1	7	40	0.646168E−07
5	10	59	3360	0.01	1	7	46	0.859645E−07
100	5	38	25561	0.01	1	6	28	0.239885E−07
5	100	35	17744	0.01	1	6	26	0.629401E−07
100	100	33	335477	0.15	1	6	26	0.499880E−07

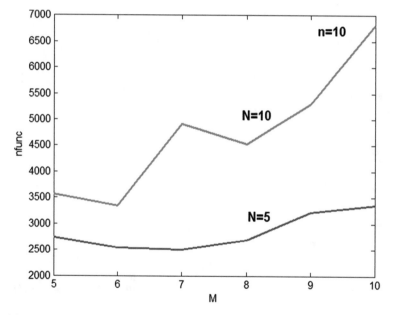

Fig. 2.1 Evolution of *nfunc*, $n = 10$

number of evaluations of the minimizing function, and *cpu* is the CPU time in
seconds.

Figure 2.1 shows the evolution of the number of the function evaluations (*nfunc*)
subject to $M = 5, 6, 7, 8, 9, 10$ for two different values of N: $N = 5, 10$.

Table 2.2 shows the performances of DEEPS for different values of N and M,
where this time, $n = 100$, with the same bounds of the domains D and $D_j, j = 1, \ldots, N$.

Figure 2.2 presents the evolution of *nfunc* subject to $M = 10, 20, 30, 40, 50, 60,
70, 80, 90, 100$ for two different values of N: $N = 10, 20$. Observe that there are some
combinations of N and M for which the performances of DEEPS are the best. In this

Table 2.2 Performances of DEEPS for minimizing the function VARDIM
$n = 100$, $f(x_0) = 0.131058E + 15$

N	M	iter	nfunc	cpu	BR	BCR	CR	$f(x^*)$
10	10	13153	1517516	6.52	1	16	11577	0.154579E−04
50	50	287	743570	3.28	1	10	230	0.128920E−04
10	100	360	366038	1.63	1	10	296	0.119802E−04
100	10	377	449312	1.85	1	11	318	0.119262E−04
50	100	218	1110218	4.82	1	10	186	0.153634E−04
100	50	243	1260896	5.45	1	10	190	0.140270E−04
100	100	205	2088811	9.04	1	10	167	0.146979E−04
1000	100	135	13759103	59.22	1	9	109	0.177911E−04
100	1000	124	12423245	59.17	1	9	100	0.133474E−04

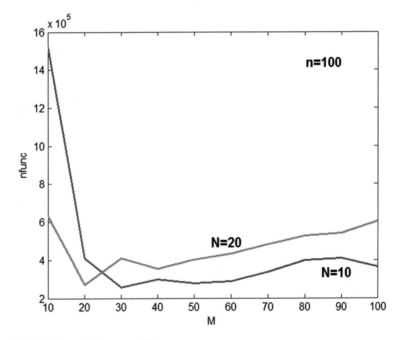

Fig. 2.2 Evolution of *nfunc*, $n = 100$

case, for $n = 100$, the best performances are obtained for $N = 10$ and $M = 30$, or for $N = 20$ and $M = 20$.

Some comments are as follows:

1. DEEPS is able for solving relatively large-scale unconstrained optimization problems in a reasonable amount of time. Observe that for any values of N and M, DEEPS finds a point x^* such that $f(x^*) \leq f(x_0)$. Nothing can be said about the optimality of x^* (see Remark 1.1).

2. For larger values of N and M, DEEPS needs more function evaluations and more computing time, but fewer iterations in order to get the solution. For example, for $n = 100$ with $N = 1000$ and $M = 1000$, DEEPS needs 108 iterations, but 108207523 evaluations of the minimizing function and 693.86 seconds to get a minimum point.
3. Better results are obtained when N and M are smaller or equal to n and $N < M$.
4. *For a given problem, choosing the best values of N and M is an open question.*

2.3.2 The Choice of Domains D and D_j, $j = 1, \ldots, N$.

Often, the real unconstrained optimization problems have some bounds on variables. These bounds are defined by constructive limitations of the physical phenomenon represented by the minimization problem. However, there are problems in which the variables are not bounded. In our approach, we consider that the variables are equally lower and upper bounded by the same values, ad hoc introduced, and the minimum point is sought around the initial point x_0 in the domain defined by these bounds. Therefore, the domain D is defined by some lower and upper bounds, *lobnd* and *upbnd*, the same for all variables. On the other hand, the local domains $D_j, j = 1, \ldots, N$, are defined by some lower and upper bounds, *lobndc* and *upbndc*, smaller than the bounds defining the domain D. At some specific iteration (see steps: 14 and 16 of DEEPS), the bounds defining these domains are reduced, thus narrowing the area of research for the minimum point.

In our numerical experiments, we noticed that the bounds associated to these domains are of some importance on the performances of the algorithm, but they are not crucial. For example, Table 2.1 presents the performances of DEEPS for minimizing the above VARDIM problem (2.2) with $n = 10$, with the following bounds: *lobnd* $= 0$, *upbnd* $= + 1$, *lobndc* $= -0.1$ and *upbndc* $= +0.1$. Now, if the bounds are changed as *lobnd* $= 0$, *upbnd* $= + 10$, *lobndc* $= -0.5$ and *upbndc* $= +0.5$, then, when $n = 10$, for some values of N and M, the performances of DEEPS are as in Table 2.3.

Comparing these performances of DEEPS from Table 2.3 with those from Table 2.1, we see that there are some differences, but not significant. In other words, the bounds on the variables defining the domains D and $D_j, j = 1, \ldots N$, are not crucial.

Table 2.4 presents the performances of DEEPS with $n = 100$, *lobnd* $= 0$, *upbnd* $= + 10$, *lobndc* $= -0.5$ and *upbndc* $= +0.5$, for some values of N and M.

Comparing Tables 2.2 and 2.4, we see that by enlarging the bounds of the domains D and $D_j, j = 1, \ldots N$, the DEEPS algorithm needs more iterations, more evaluations of the minimizing function, and, therefore, more computing time.

Table 2.3 Performances of DEEPS for minimizing the function VARDIM
$n = 10, f(x_0) = 0.219855E + 07$

N	M	iter	nfunc	cpu	BR	BCR	CR	$f(x^*)$
3	3	501	6306	0.01	3	13	497	0.885741E−07
5	2	444	7545	0.01	3	13	439	0.563607E−07
5	5	66	2161	0.00	3	10	62	0.773045E−07
10	5	61	4100	0.00	3	9	53	0.537187E−07
5	10	55	3169	0.00	3	9	48	0.671316E−07
5	100	34	17248	0.01	2	9	30	0.423977E−07
100	5	37	25405	0.01	3	9	33	0.862972E−07
100	100	31	315375	0.13	3	9	27	0.589694E−07

Table 2.4 Performances of DEEPS for minimizing the function VARDIM
$n = 100, f(x_0) = 0.131058E + 15$

N	M	iter	nfunc	cpu	BR	BCR	CR	$f(x^*)$
50	50	381	988144	4.28	4	12	282	0.143439E−04
50	100	301	1533231	6.58	4	12	245	0.133380E−04
100	50	247	1282335	5.53	4	12	197	0.166122E−04
100	10	546	651009	2.70	4	13	403	0.139601E−04

Some comments are in order:

1. For different bounds *lobnd*, *upbnd*, *lobndc* and *upbndc* the DEEPS algorithm gives a solution x^* to the optimization problem, for which $f(x^*) \leq f(x_0)$.
2. Normally, the optimization algorithms and especially the derivative-free optimization ones must be assisted by the user subject to the process of tailoring the parameters associated to them, in our case, the bounds of the domains where the minimum is sought. For larger values of bounds *lobnd*, *upbnd*, *lobndc* and *upbndc* DEEPS needs more function evaluations and, therefore, more computing time.
3. *For a given problem, choosing the best values of the bounds: lobnd, upbnd, lobndc and upbndc is an open problem.* The best results are obtained for reasonably small domains around the initial point, emphasizing once again the local character of the algorithm as we can see in the next subsection.

2.3.3   Local versus Global Search

Around a specified initial point x_0, the algorithm determines a point, for which the function value might be smaller than $f(x_0)$. Therefore, the searching for the minimum point is local. Finding a global minimum point depends on the number of the points randomly chosen in the domains D and $D_j, j = 1, \ldots, N$, and on the bounds of these domains, but more importantly, it depends on the properties and characteristics of the minimizing function.

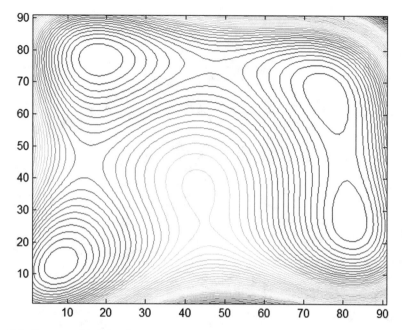

Fig. 2.3 Contours of function f_1

For example, let us minimize the function

$$f_1(x) = (x_1^2 + x_2 - 11)^2 + (x_1 + x_2^2 - 7)^2 \tag{2.3}$$

(Himmelblau 1972). The contour of this function is represented as in Fig. 2.3.

This function has 4 minimum points, and for each of them, the function value is zero. Considering $N = 50$, $M = 100$, $lobnd = -2$, $upbnd = +2$, $lobndc = -0.1$, $upbndc = +0.1$ and $\varepsilon = 10^{-6}$, then, for different initial points, the DEEPS algorithm gives different minimum points. For $x_0 = [1, \ 1]$, the minimum point generated by DEEPS is $x_1^* = [3, \ 2]$, for which $f_1(x_1^*) = 0$. For $x_0 = [-1, \ -1]$, the solution is $x_2^* = [-3.7792, \ -3.2830]$, where $f_1(x_2^*) = 0$. For $x_0 = [-5, \ 3]$, the solution is $x_3^* = [-2.8051, \ 3.1313]$, where $f_1(x_3^*) = 0$. Again, for $x_0 = [5, \ -3]$, the solution is $x_4^* = [3.5844, \ -1.8481]$, where $f_1(x_4^*) = 0$. Observe that the global minimum is zero, obtained for different minimum points.

Another example is more illustrative. Consider the function

$$f_2(x) = 3(1 - x_1)^2 \exp\left(-x_1^2 - (x_2 + 1)^2\right)$$
$$- 10\left(x_1/5 - x_1^3 - x_2^5\right) \exp\left(-x_1^2 - x_2^2\right) - \exp\left(-(x_1 + 1)^2 - x_2^2\right)/3. \tag{2.4}$$

The surface plot and the contour plot of this function are represented in Fig. 2.4.

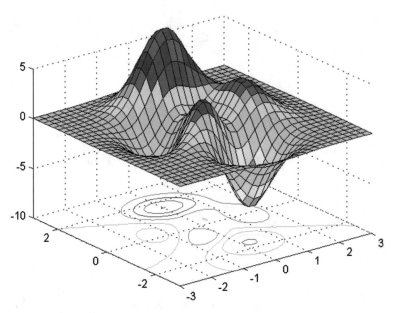

Fig. 2.4 The surface plot and the contour plot of $f_2(x)$

For the initial point $x_0 = [1, \ 1]$, $f_2(x_0) = 2.4337$ and DEEPS give the solution $x_1^* = [-1.5311, \ 0.1415]$, for which $f_2(x_1^*) = -3.5308$. On the other hand, for $x_0 = [3, \ -2]$, $f_2(x_0) = 0.000416$ and the solution is $x_2^* = [0.2287, \ -1.6259]$, where $f_2(x_2^*) = -6.5479$. Obviously, DEEPS has a local character, that is, for different initial points x_0, it finds different points x^*, for which $f_2(x^*) \leq f_2(x_0)$.

2.3.4   The Stopping Criterion

We have assumed in this algorithm that a minimum value of the function f is known as f^{opt}. Therefore, the stopping criterion used in step 16 is based on this known minimum value of the function f. However, some other criteria for stopping the algorithm may be used. For example, after a specified number of iterations or after a specified number of function evaluations, the iterations are stopped. For the numerical experiments presented in this work, we use the criterion based on f^{opt}. Observe that, since the points in which the function is evaluated are randomly generated, then, close to the solution, the difference between the values of the minimizing function computed into the minimum and the maximum points, respectively, might be larger than a small enough prespecified value. This is the reason why this difference cannot be used as a stopping criterion. For a given function f, the minimum point is not known. In this case, the algorithm is run with a prespecified value for f^{opt}, let us say $f^{opt} = 0$, until an estimation of the minimum value is obtained. With this value of the minimum, the algorithm is run again, allowing its performances to be seen.

2.3.5 Line Search in DEEPS

Intensive numerical experiments have shown that the optimization process of DEEPS has two distinct phases: the *reduction phase,* in which the function values are strongly reduced, and the *stalling phase,* in which the reduction of the function values is very slow. Besides, the number of iterations in the stalling phase is substantially larger than the corresponding number of iterations in the reduction phase. In order to reduce the stalling phase in step 12 of the algorithm DEEPS, a simple line search procedure is introduced. However, this does not apply to every optimization problems. In other words, for some optimization problems, this line search included in step 12 of the algorithm is beneficial, thus reducing the optimization process. For some other optimization problems, this line search is not useful. Defining a criterion according to which the line search in step 12 should be used is an open problem.

In step 12 of the algorithm DEEPS, if $\left| f_{iter}^{min} - f_{iter-1}^{min} \right| \leq \delta$, where δ is a small, positive threshold (usually $\delta = 10^{-4}$), a simple line search is initiated from the current minimum point z^{min} along a search direction determined by this minimum point and the second minimum point z^{mins}. Firstly, the point z^{min} is moved to a positive direction by computing the point $z^s = z^{min} + \beta(z^{min} - z^{mins})$, where β is a positive step (usually $\beta = 0.01$). If the value of the minimizing function in this new point is smaller than the value of the function in the current minimum point, then set $z^{min} = z^s$ and continue the algorithm. Otherwise, the point z^{min} is moved to a negative direction by computing the point $z^s = z^{min} - \beta(z^{min} - z^{mins})$. Again, if the value of the minimizing function in this new point is smaller than the value of the function in the current minimum point, then set $z^{min} = z^s$ and continue the algorithm. The number of iterations with successful line searches of the above procedure is denoted as LS. The idea of this line search is that in the stalling phase, for example, the iterations enter into a narrow valley, where the function values are reducing very slowly. Therefore, a procedure to accelerate the convergence of the algorithm is to implement a line search from the current minimum point along a search direction determined by the minimum point and the second minimum point. For some optimization problems, this idea proves to be useful.

Let us consider two examples to illustrate the efficiency of the line search from step 12 of the algorithm DEEPS. Consider the function CUBE (Bongartz et al. 1995):

$$f(x) = (x_1 - 1)^2 + \sum_{i=2}^{n} 100(x_i - x_{i-1}^3)^2, \tag{2.5}$$

with $n = 4$ and the initial point $x_0 = [-1.2, 1, \ldots, -1.2, 1]$, for which $f(x_0) = 1977.237$. Considering $lobnd = -5$, $upbnd = +5$, $lobndc = -1$ and $upbndc = +1$, with $\varepsilon = 10^{-4}$, then the minimum point determined by DEEPS is $x^* = [1, 1, 1, 1]$, where $f(x^*) = 0$. Observe that $f(x^*) < f(x_0)$. The performances of DEEPS with line search (with step 12) and without line search (with step

Table 2.5 Performances of DEEPS for minimizing the function CUBE

	With line search	Without line search
iter	1735	2838
nfunc	44990111	73765757
cpu	7.91	17.76
CR	1734	2837
BR	0	0
BCR	13	13
LS	1695	0

12 inhibited) are presented in Table 2.5, where LS is the number of successful line search, and *cpu* is the CPU computing time in seconds.

Observe that the line search introduced in step 12 of the algorithm is effective: the number of iterations with successful line searches (1695) is close to the total number of iterations (1735). The evolution of the values of the minimizing function along the iterations with line search and without line search is shown in Fig. 2.5.

From Fig. 2.5, in the case of DEEPS with line search, we see that in the first 214 iterations (i.e., 12.33% of the total number of iterations), the values of the minimizing function CUBE are reduced from 1977.2370 to 0.00098. For the rest of 1521 iterations (i.e., 87.67%), the function values are reduced from 0.00098 to 0.9870E-04. In the case of DEEPS without line search, again, in the first 251 iterations (i.e., 8.84%), the values of the CUBE function are reduced from 1977.2370 to 0.000998. For the rest of 2587 iterations (i.e., 91.16%), the function values are reduced from 0.000998 to 0.000096. This is the typical evolution of the function values given by DEEPS.

Now, let us consider the function ARWHEAD (Bongartz et al. 1995):

$$f(x) = \sum_{i=1}^{n-1} (-4x_i + 3) + \sum_{i=1}^{n-1} \left(x_i^2 + x_n^2 \right)^2, \tag{2.6}$$

with $n = 40$ and the initial point: $x_0 = [1, 1, \ldots, 1]$, for which $f(x_0) = 117.0$. Considering $lobnd = -50$, $upbnd = +50$, $lobndc = -10$ and $upbndc = +10$, with $\varepsilon = 10^{-5}$, then the minimum point determined by DEEPS is $x^* = [1, 1, \ldots, 1, 0]$, for which $f(x^*) = 0$. The performances of DEEPS with line search (with step 12) and without line search (with step 12 inhibited) are presented in Table 2.6. Observe that the DEEPS algorithm with line search needs 151 iterations to get the point x^*, while without line search, it needs 69 iterations. For this function, the bounds *lobnd* and *upbnd* are twice reduced, and the local bounds *lobndc* and *upbndc* are 15 times reduced. The performances of the DEEPS algorithm without line search are obviously better.

The evolution of the values of the minimizing function along the iterations with line search and without line search, respectively, is presented in Fig. 2.6. For the DEEPS algorithm without line search, in the first 52 iterations (i.e., 75.36%), the function values are reduced from 117.0 to 0.000944. For the rest of 17 iterations (i.e., 24.64%), the function values are reduced to 0.0000022.

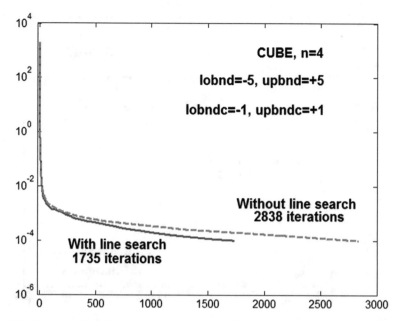

Fig. 2.5 Evolution of the function values

Table 2.6 Performances of DEEPS for minimizing the function ARWHEAD

	With line search	Without line search
iter	151	69
nfunc	391195	178925
cpu	1.39	0.55
CR	148	66
BR	2	2
BCR	15	15
LS	97	0

In the first 82 iterations (i.e., 54.30%), the DEEPS algorithm with line search reduces the function values from 117 to 0.00093. For the rest of 69 iterations (i.e., 45.70%), the function values are reduced to 0.0000016. Observe that the stalling phase for the DEEPS algorithm with line search is longer than the same phase for the DEEPS algorithm without line search. *Anyway, using or not using the line search in the DEEPS algorithm is an open problem.* The recommendation is to run DEEPS without line search and to see its performances. After that, rerun DEEPS with line search.

Figures 2.5 and 2.6 present two typical evolutions of the function values along the iterations of the DEEPS algorithm. In the first one, the function values are suddenly reduced for a small number of iterations. In the second one, the function values are steadily reduced, having some plateaus, where the function values are relatively

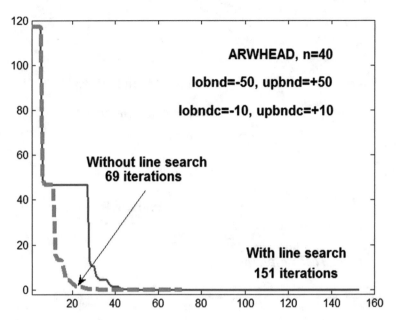

Fig. 2.6 Evolution of the function values

constant along the iterations, which correspond to the halving of the bounds of the domains, where the trial points or the local trial points are generated, or to the reduction of the complex. More details are given in Chap. 4.

References

Bongartz, I., Conn, A.R., Gould, N.I.M., & Toint, Ph.L (1995). CUTE: constrained and unconstrained testing environments. ACM Transactions on Mathematical Software, 21, 123-160.

Himmelblau, D.M., (1972). *Applied Nonlinear Programming*. McGraw-Hill, New York.

Chapter 3
Convergence of the Algorithm

Abstract The purpose of this chapter is to prove the convergence of the algorithm. It is shown that the evolution of the maximum distance among the trial points and the local trial points tend to zero. For continuous and lower bounded functions, the two-level random search algorithm is convergent to a point for which the function value is smaller or equal to the value of the minimizing function in the initial point.

Keywords Maximum distance among the trial points · Maximum distance among the trial points and the local trial points

Suppose that function f is continuous and bounded below on $\mathbb{R}^n$. The algorithm follows a strategy for seeking a minimum of this function, based on generating some points in specified domains around the initial point x_0 (Andrei 2020). The key element of the algorithm is the procedure for selecting the best point around every trial point in the domain $D_j, j = 1, \ldots, N$. At each iteration, a new set of trial points is produced, for which the values of the minimizing function f are smaller. The rationale behind this approach is that at every iteration, the best (minimum) points are always chosen and possibly, the complex defined by the trial points is reduced, and the domain D and the local domains $D_j, j = 1, \ldots, N$, shrink by halving their bounds, thus narrowing the area of searching. Therefore, the algorithm generates a sequence of points $\{x_k\}$, for which the corresponding sequence $\{f(x_k)\}$ is monotonously decreasing and bounded. It follows that the sequence $\{f(x_k)\}$ is convergent to a value $f(x^*)$. Therefore, by continuity, the corresponding sequence $\{x_k\}$ is convergent to a point x^*, for which $f(x^*) \leq f(x_0)$. Nothing can be said about the optimality of x^*. We can only say that the algorithm determines a point x^*, for which $f(x^*) \leq f(x_0)$, see Remark 1.1. In the following, let us discuss and present some key elements of the algorithm by illustrating the evolution of the *maximum distance among the trial points*, as well as the evolution of the *maximum distance among the trial points and the local trial points* along the iterations.

N. Andrei, *A Derivative-free Two Level Random Search Method for Unconstrained Optimization*, SpringerBriefs in Optimization,
https://doi.org/10.1007/978-3-030-68517-1_3

3.1   Evolution of the Maximum Distance Among the Trial Points

At iteration *iter*, for any two points x_i and x_j, from the set of trials points, let us define the distance between these points as $d_{ij}^{iter} = \left\| x_i - x_j \right\|_2$, for $i, j = 1, \ldots, N, i \neq j$. Of course, this distance depends on the bounds *lobnd* and *upbnd*. Now, let us determine a pair of trial points x_u and x_v, from the set of trials points so that

$$d_{uv}^{iter} = \left\| x_u - x_v \right\|_2 = \max_{i,j=1,\ldots,N, j>i} d_{ij}^{iter}, \tag{3.1}$$

that is, d_{uv}^{iter} is the maximum distance among the trial points. With this, define the sphere S^{iter} centered in point $c_{uv} = (x_u + x_v)/2$ of ray $r_{uv}^{iter} = d_{uv}^{iter}/2$. Obviously, the volume of this sphere is

$$V^{iter} = \frac{\pi}{6} \left(d_{uv}^{iter} \right)^3. \tag{3.2}$$

Proposition 3.1 *At iteration iter, all the trial points belong to the sphere S^{iter}.*

Proof Suppose by contradiction that a trial point $x_p \notin S^{iter}$. Then, if $\left\| x_u - x_p \right\| \leq \left\| x_u - x_v \right\|$, then $\left\| x_v - x_p \right\| \geq \left\| x_u - x_v \right\|$, that is, d_{uv}^{iter} is not the maximum distance among the trail points at iteration *iter*. Similarly, if $\left\| x_v - x_p \right\| \leq \left\| x_u - x_v \right\|$, then $\left\| x_u - x_p \right\| \geq \left\| x_u - x_v \right\|$, that is, d_{uv}^{iter} is not the maximum distance among the trail points at iteration *iter*. Therefore, x_p has to belong to S^{iter}. ♦

Proposition 3.2 *At iteration iter, if $\left| f_{iter}^{min} - f_{iter-1}^{min} \right| < t$, then $V^{iter} = V^{iter-1}/8$.*

Proof Let d_{uv}^{iter-1} be the maximum distance among the trial points at iteration *iter* − 1 determined by the trial points x_u and x_v. Observe that at iteration *iter*, if $\left| f_{iter}^{min} - f_{iter-1}^{min} \right| < t$, then the algorithm DEEPS reduces the complex, that is, for $j = 1, \ldots, N$, the new trial points are computed as $x_j = x^{min} + (x_j - x^{min})/2$, where x^{min} is the trial point corresponding to the minimum value of function *f*. Therefore, the distance from any trial point x_i to any trial point x_j is

$$d_{ij}^{iter} = \left\| x_i - x_j \right\|_2 = \left\| x^{min} + (x_i - x^{min})/2 - x^{min} - (x_j - x^{min})/2 \right\|_2$$
$$= \frac{1}{2} \left\| x_i - x_j \right\|_2 = \frac{1}{2} d_{ij}^{iter-1}. \tag{3.3}$$

In particular, for the points x_u and x_v, which define the maximum distance among the trial points at iteration *iter*, it follows that $d_{uv}^{iter} = d_{uv}^{iter-1}/2$. Therefore, from (3.2), $V^{iter} = V^{iter-1}/8$. ♦

Hence, at iteration *iter*, if $\left| f_{iter}^{min} - f_{iter-1}^{min} \right| < t$, then the complex is reduced, and the DEEPS algorithm generates the trial points in a sphere of smaller volume.

Now, let us consider the situation in which at a certain iteration $iter - 1$, around the point x_0, a number N of trial points are generated as

$$x_j^{iter-1} = x_0 + r_j(upbnd - lobnd) + e(lobnd), \quad j = 1, \ldots, N,$$

where: $r_j = [r_j^1, \ldots, r_j^n]^T$ is a vector with the components r_j^k, $k = 1, \ldots, n$, random numbers in $[0, 1)$; $e = [1, \ldots, 1] \in \mathbb{R}^n$, and $lobnd$ and $upbnd$ are the bounds of the domain D, (see step 2). As we know, the distance among these trial points is

$$d_{ij}^{iter-1} = \left\| x_i^{iter-1} - x_j^{iter-1} \right\|_2 = \|r_i - r_j\|_2 |upbnd - lobnd|,$$
$$i, j = 1, \ldots, N, \quad i \neq j. \tag{3.4}$$

But, $\|r_i - r_j\|_2 = \left(\sum_{k=1}^n (r_i^k - r_j^k)^2 \right)^{1/2} \leq \sqrt{n}$. Therefore, $d_{ij}^{iter-1} \leq \sqrt{n} |upbnd - lobnd|$, for any $i, j = 1, \ldots, N, i \neq j$.

Proposition 3.3 *If at the iteration iter $\left| f_{iter}^{max} \right| > B$, where B is a large enough prespecified parameter, then, at this iteration, the distance among the trial points is reduced.*

Proof In this case, at step 14, the algorithm DEEPS reduces the bounds $lobnd$ and $upbnd$ as $lobnd/2$ and $upbnd/2$, respectively, and go to generate a new set of trial points in step 2 using these new bounds. Hence, at this iteration, the new trial points are computed as $x_j^{iter} = x_0 + \frac{1}{2} r_j(upbnd - lobnd) + \frac{1}{2} e(lobnd)$, for $j = 1, \ldots, N$, where again r_j is a vector with components random numbers in $[0, 1)$. Therefore, at this iteration, the distance among the new trial points is

$$d_{ij}^{iter} = \left\| x_i^{iter} - x_j^{iter} \right\|_2 = \frac{1}{2} \|r_i - r_j\|_2 |upbnd - lobnd| = \frac{1}{2} d_{ij}^{iter-1}, \tag{3.5}$$

for any $i, j = 1, \ldots, N, i \neq j$. ♦

For example, for function VARDIM with $n = 10$, $\varepsilon = 10^{-5}$, for different values of $lobnd$ and $upbnd$, the evolution of the maximum distance d_{uv}^{iter}, $iter = 1, 2, \ldots$, among the trial points is as in Fig. 3.1.

Observe that the maximum distance among the trial points tends to zero. This is a result of the reduction of the complex and of course the reduction of the bounds $lobnd$, $upbnd$ as well as of the reduction of the local bounds $lobndc$ and $upbndc$. In other words, along the iterations, the algorithm generates the trial points in smaller and smaller domains. Besides, subject to the number of iterations or to the number of function evaluations, observe that the value of the bounds $lobnd$ and $upbnd$ is not critical.

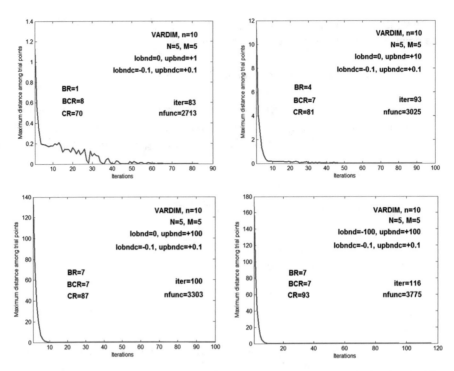

Fig. 3.1 Maximum distance among the trial points for different bounds *lobnd* and *upbnd* for problem VARDIM

For the same function, for different values of *lobnd* and *upbnd*, the performances of DEEPS are similar. Observe that for larger bounds of the domain D, the values of the minimizing function have a very broad reduction in the first iterations. For example, for *lobnd* = 0 and *upbnd* = + 100 (see Fig. 3.1), in the first 23 iterations (i.e., 23% from the total number of iterations), the maximum distance among the trial points is reduced from 139.4926 to 0.1061. For the rest of 77 iterations (i.e., 77%), the values of this distance are reduced to 0.5158E−6. This is typical behavior of DEEPS. It is worth mentioning that the performances of DEEPS are very little modified for larger values of the bounds of the domain D. For example, for *lobnd* = − 100 and *upbnd* = + 100, the same solution is obtained in 116 iterations and 3775 evaluations of the minimizing function.

3.2   Evolution of the Maximum Distance Among the Trial Points and the Local Trial Points

If $\left| f_{iter}^{min} - f^{opt} \right|_{iter} = \left| f_{iter-1}^{min} - f^{opt} \right|_{iter-1}$, then, at step 16 of the algorithm DEEPS, the local bounds *lobndc* and *upbndc* are halved. Now, at iteration *iter*, for every trail point $x_j, j = 1, \ldots, N$, let us define the distance $d_j^{iter}, j = 1, \ldots, N$, from $x_j, j = 1, \ldots, N$, to the local points $x_j^k, k = 1, \ldots M$, generated by the algorithm, that is,

$$d_j^{iter} = \max_{k=1, \ldots, M} \left\| x_j - x_j^k \right\|_2, \quad j = 1, \ldots N. \tag{3.6}$$

With this, at iteration *iter*, let us determine a trial point x_w, for which

$$d_w^{iter} = \max \left\{ d_j^{iter} : j = 1, \ldots, N \right\}. \tag{3.7}$$

In other words, d_w^{iter} is the maximum distance among all trial points and all local trial points. As in Proposition 3.3, the following proposition can be proved.

Proposition 3.4 *If at the iteration iter* $\left| f_{iter}^{min} - f^{opt} \right|_{iter} = \left| f_{iter-1}^{min} - f^{opt} \right|_{iter-1}$ *holds, then, at this iteration, the distance d_w^{iter} among the trial points and the local trial points is reduced.*

Again, since at every iteration the algorithm DEEPS chooses the minimum point among the generated local points, and at some specific iteration, the bounds *lobndc* and *upbndc* are halved, it follows that the maximum distance between the trial points and the local trial points tends to zero. For example, for the function VARDIM with $n = 10$, $\varepsilon = 10^{-5}$, the evolution of d_w^{iter} along the iterations for different values of *lobndc* and *upbndc* is as in Fig. 3.2.

With these, the convergence of the algorithm DEEPS can be proved as follows:

Theorem 3.1 *Let f be a continuous function bounded from below. Then, the algorithm DEEPS initialized in a point x_0 generates a sequence $\{x_k\}$ convergent to a point x^*, where $f(x^*) \leq f(x_0)$.*

Proof Since at every iteration k DEEPS chooses the minimum point x_k from the trial points, it follows that the sequence $\{f(x_k)\}$ is monotonously decreasing, that is, $f(x_k) \leq f(x_{k-1})$, for any $k = 1, 2, \ldots$ Function f is bounded from below. Therefore, the sequence $\{f(x_k)\}$ is convergent. By continuity, it follows that there is a point x^* such that $\lim_{k \to \infty} f(x_k) = f(\lim_{k \to \infty} x_k) = f(x^*)$. ◆

Observe that x^* is only a point where $f(x^*) \leq f(x_0)$. Since we do not have access to the gradient (and the Hessian) of function f, nothing can be said about its optimality. Having in view the derivative scarcity of information about the minimizing function f, the result given by DEEPS may be acceptable from a practical point of view.

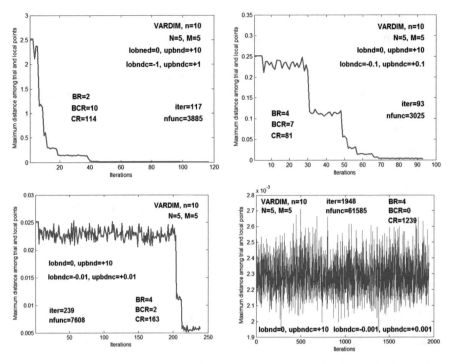

Fig. 3.2 Maximum distance among the trial points and the local trial points for different local bounds *lobndc* and *upbndc* for the problem VARDIM

Reference

Andrei, N., (2020). *A simple deep random search method for unconstrained optimization. Preliminary computational results.* Technical Report No.1/2020. AOSR – Academy of Romanian Scientists, Bucharest, Romania, February 29, 2020. (87 pages)

Chapter 4
Numerical Results

Abstract Some numerical results with an implementation of the DEEPS algorithm, as well as the evolution of the minimizing function values, along the iterations of the optimization process are presented in this chapter. It is shown that the optimization process has two phases. In the first one, the reduction phase, for a small number of iterations, the function values are strongly reduced. In the second one, the stalling phase, the function values are very slowly reduced for a large number of iterations. For some problems, in the first phase of the optimization process, the function values are steadily reduced with some plateaus corresponding to halving the bounds of the domains where the trial points are generated. This is a typical evolution of the function values along the iterations of DEEPS. Intensive numerical results for solving a number of 140 unconstrained optimization problems, out of which 16 are real applications, have proved that DEEPS is able to solve a large variety of problems up to 500 variables. Comparisons with the Nelder-Mead algorithm showed that DEEPS is more efficient and more robust.

Keywords Narrow positive cone function · Modified Wolfe function · Large-scale problems · Performance profile

In the following, let us report the performances of DEEPS for solving some unconstrained optimization problems. Firstly, two unconstrained optimization problems are considered, for which the known derivative unconstrained optimization methods risk to terminate at a saddle point. Secondly, to get some insights into the behavior of DEEPS, the evolution of the function values along the iterations is presented for a number of six functions. Next, the performances of DEEPS for solving 140 unconstrained optimization problems, out of which 16 are real applications, are compared versus the Nelder-Mead algorithm. Finally, the performances of DEEPS for solving 30 large-scale unconstrained optimization problems up to 500 variables are shown.

4.1   Two Unconstrained Optimization Problems

The first is a modification given by Frimannslund and Steihaug (2011) of a problem
suggested by Wolfe (1971). In its unmodified form, this problem was used to show
that the gradient methods tend to converge to a saddle point. The modification makes
the function bounded below and introduces a local minimum. The second problem is
a modification, also given by Frimannslund and Steihaug (2011), of a problem
presented by Abramson (2005). The modified problem has a very narrow cone of
negative curvature, being bounded below with some local minimizers. An interesting discussion on the performances of the methods CSS-CI (see Frimannslund and
Steihaug 2007), NEWUOA (see Powell 2004), and NMSMAX (see Higham n.d.)
for solving these problems with two variables is given by Frimannslund and
Steihaug (2011). In the following, let us present the performances of DEEPS for
solving these problems with different numbers of variables.

Problem 1 (Modified Wolfe function)

$$f_1(x) = \sum_{i=1}^{n/2} \frac{1}{3}x_{2i-1}^3 + \frac{1}{2}x_{2i}^2 - \frac{2}{3}\left(\min\left\{x_{2i-1}, -1\right\} + 1\right)^3.$$

Initial point: $x_0 = [1, \ldots 1]$. Figure 4.1 shows the graphical representation of the
modified Wolfe function for $n = 2$.

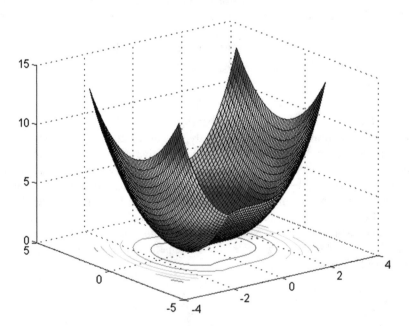

Fig. 4.1 Modified Wolfe function, $n = 2$

Table 4.1 Performances of DEEPS for solving Problem 1, $\varepsilon = 0.0001$

n	N	M	iter	nfunc	cpu	CR	BR	BCR	f^*	$f(x_0)$
2	5	2	32	569	0.01	31	0	4	−3.885618	0.833333
20	5	20	84	9060	0.04	83	0	11	−38.85618	8.333333
40	100	500	127	6374286	22.82	126	0	12	−77.712361	16.66666
50	50	100	744	3779615	12.26	743	0	12	−97.140451	20.83333

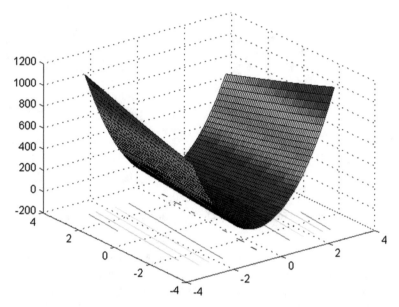

Fig. 4.2 A narrow positive cone, $n = 2$

For different values of the number of variables, the performances of DEEPS are as in Table 4.1.

Problem 2 (A narrow positive cone)

$$f_2(x) = \sum_{i=1}^{n/2} (9x_{2i-1} - x_{2i})(11x_{2i-1} - x_{2i}) + \frac{1}{2}x_{2i-1}^4,$$

Initial point: $x_0 = [2, \ldots 2]$. The graphical representation of the narrow positive cone for $n = 2$ is presented in Fig. 4.2.

For different values of the number of variables, the performances of DEEPS for solving this problem are as in Table 4.2.

In Tables 4.1 and 4.2, *iter* is the number of iterations, *nfunc* is the number of function evaluations, *cpu* is the CPU time in seconds, CR is the number of complex reductions, BR is the number of reductions of the bounds *lobnd* and *upbnd*, and BCR is the number of reductions of the local bounds *lobndc* and *upbndc*. f^* is the value of the minimizing function in the minimum point.

Table 4.2 Performances of DEEPS for solving Problem 2, $\varepsilon = 0.00001$

n	N	M	$iter$	$nfunc$	cpu	CR	BR	BCR	f^*	$f(x_0)$
2	5	50	488	125093	0.26	486	0	2	−0.5	328.0
4	50	100	325	1654925	0.81	324	0	2	−1.0	656.0
8	1000	500	995	499444169	344.06	994	0	5	−2.0	1312.0
10	500	500	3648	915497674	765.90	3647	0	7	−2.5	1640.0

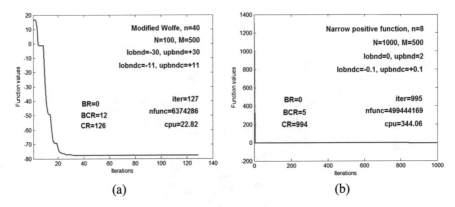

Fig. 4.3 Function evolution for modified Wolfe and for narrow positive cone functions

For these two functions, Fig. 4.3 presents the evolution of the function values along the iterations. For both these functions, DEEPS monotonically reduces the function values. Observe that the algorithm works very well in practice. In the reduction phase, the algorithm produces a rapid decrease in the function values. After that, the algorithm enters in the stalling phase. This phase is more emphasized for the narrow positive cone function, where, as it can be seen, the function values decrease very slowly.

From Fig. 4.3(a), we can see that, for the modified Wolfe function $f_1(x)$ with $n = 40$, the function values are reduced along the iterations, but there are some *plateaus*, where the algorithm is stalling for a small number of iterations, that is, it reduces the function value very slowly. For this function, the abrupt reductions of the function values appear at the iterations, where the complex is reduced, or where the bounds *lobndc* and *upbndc* are halved. In this case, CR = 126, BR = 0, and BCR = 12. This is a typical evolution of the DEEPS algorithm, that is, for different bounds *lobnd*, *upbnd*, *lobndc* and *upbndc*, the reduction of the values of the minimizing function along the iterations has the same allure.

From Fig. 4.3(b), observe that the function values are drastically reduced for a small number of iterations from the very beginning of the optimization process. For the narrow positive cone function $f_2(x)$, in the first 16 iterations (i.e., 1.6% from the total number of iterations), the function values are reduced from 1312.0 to −0.00816. After that, for the rest of 979 iterations (i.e., 98.4%), the function values are very slowly reduced up to the final value −2. It is worth mentioning that, for the minimizing function $f_2(x)$ with $n = 8$, the steepest descent method needs 13741 iterations and 27483 evaluations of $f_2(x)$. The corresponding evolution of the function values along the iterations for the steepest descent method is similar to that in Fig. 4.3(b). In the first three iterations, the steepest descent method reduces the function value from 1312.0 to -0.1859. After that, for the rest of 13738 iterations, the function values are reduced very slowly to the optimal value -2. More on this behavior of DEEPS is developed in the next subsection, where the evolution of the reduction of the function values is shown for a number of six functions.

4.2 Typical Evolution of the Minimizing Function in the Optimization Process

Let us now emphasize the evolution of the minimizing function given by the algorithm DEEPS for some functions. Consider the following functions from the CUTE collection (Bongartz et al. 1995), where their algebraic expression, the number of variables, and the initial point are presented for each of them.

1. *Beale Function*:

$$f(x) = \sum_{i=1}^{n/2} \left(1.5 - x_{2i-1}(1 - x_{2i})\right)^2 + \left(2.25 - x_{2i-1}\left(1 - x_{2i}^2\right)\right)^2$$
$$+ \left(2.625 - x_{2i-1}\left(1 - x_{2i}^3\right)\right)^2,$$

with $n = 8$. Initial point: $x_0 = [1., 0.8, \ldots, 1., 0.8]$. $f(x_0) = 39.315477$.

2. *BDQRTIC Function*:

$$f(x) = \sum_{i=1}^{n-4} \left(-4x_i + 3\right)^2 + \left(x_i^2 + 2x_{i+1}^2 + 3x_{i+2}^2 + 4x_{i+3}^2 + 5x_n^2\right)^2,$$

with $n = 16$. Initial point: $x_0 = [1., 1., \ldots, 1.]$. $f(x_0) = 2712$.

3. *Trigonometric Function*:

$$f(x) = \sum_{i=1}^{n} \left(\left(n - \sum_{j=1}^{n} \cos\left(x_j\right)\right) + i(1 - \cos\left(x_i\right)) - \sin\left(x_i\right)\right)^2,$$

with $n = 20$. Initial point: $x_0 = [0.2, \ldots, 0.2]$. $f(x_0) = 3.614762$.

4. *Brown Function:*

$$f(x) = \sum_{i=1}^{n-1} \left(x_i^2\right)^{\left(x_{i+1}^2+1\right)} + \left(x_{i+1}^2\right)^{\left(x_i^2+1\right)},$$

with $n = 10$. Initial point: $x_0 = [0.1, \ldots, 0.1]$. $f(x_0) = 0.1718986$.

5. *EG2 Function:*

$$f(x) = \sum_{i=1}^{n-1} \sin\left(x_1 + x_i^2 - 1\right) + \frac{1}{2}\sin\left(x_n^2\right).$$

with $n = 50$. Initial point: $x_0 = [1., 1., \ldots, 1.]$. $f(x_0) = 41, 6528$.

6. *ARWHEAD Function:*

$$f(x) = \sum_{i=1}^{n-1}\left(-4x_i + 3\right) + \sum_{i=1}^{n-1}\left(x_i^2 + x_n^2\right)^2,$$

with $n = 50$. Initial point: $x_0 = [1., 1., \ldots, 1.]$. $f(x_0) = 147$.

Similarly as in Fig. 4.3, from Fig. 4.4, we can see that the function values steadily decrease along the iterations with some abrupt reductions at some iterations. This is the typical behavior of the DEEPS algorithm. For the function Beale, along the 280 iterations, the number of the complex reduction (see step 17) is 279, the number of the local bounds *lobndc* and *upbndc* reduction (see step 16) is 10. The bounds *lobnd* and *upbnd* are not reduced. Observe that the function values are significantly reduced at the iterations in which the local bounds *lobndc* and *upbndc* are reduced. For the function BDQRTIC, the evolution of the function values is more interesting. In this case, along 320 iterations, the number of the complex reduction is 309, the number of the bounds *lobnd* and *upbnd* reduction (see step 14) is 10, and the number of the bounds *lobndc* and *upbndc* reduction is 22. Roughly speaking, in the first 14 iterations (i.e., 4.37% from the total number of iterations), the algorithm DEEPS reduces the function values from 2712.0 to 706.748. For the rest of 306 iterations (i.e., 95.63%), the function values are reduced from 92.124 to 42.2979, which is the minimum value. For the trigonometric function, the number of the complex reduction is 9, and the number of the bounds *lobndc* and *upbndc* reduction is 2. The bounds *lobnd* and *upbnd* are not reduced. For the Brown function, the number of the complex reduction is 9, and the number of the bounds *lobndc* and *upbndc* reduction is 0. Evolution of the values of the function EG2 along the iterations is typical of the DEEPS algorithm. In this case, the complex of the trial points was 32 times reduced, and the bounds of the local domains were 11 times reduced. Also for the ARWHEAD function, the complex of trial points was 189 times reduced, and the local bounds were 19 times reduced. In Fig. 4.4, we can see that there are some *plateaus,* where the evolution of the minimizing function is stalling, that is, it reduces very slowly. This is the typical behavior of the linear convergence.

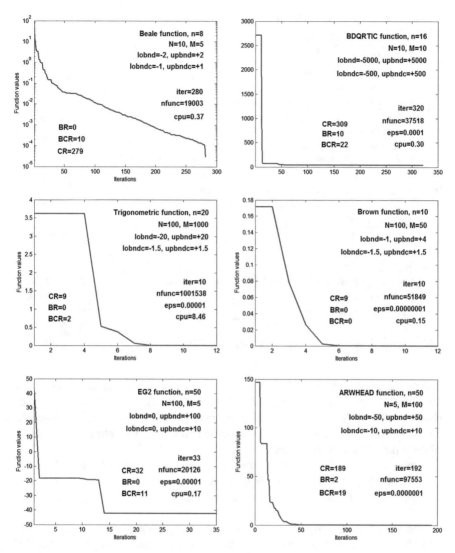

Fig. 4.4 Evolution of the function values for Beale, BDQRTIC, Trigonometric, Brown, EG2, and ARWHEAD functions

4.3   Performances of DEEPS for Solving 140 Unconstrained Optimization Problems

In this subsection, let us present the performances of DEEPS for solving 140 unconstrained optimization problems, out of which 16 are very well-known unconstrained optimization applications from different areas of activity. Table 4.3 contains the name of the applications, the initial point, the number of variables (n),

Table 4.3 Applications solved by DEEPS. Initial point. Number of variables. Accuracy

No.	Applications/initial point	n	ε
1.	Weber(1), $x_0 = [10, 10]$	2	0.00001
2.	Weber(2), $x_0 = [1.2, 1]$	2	0.000001
3.	Weber(3), $x_0 = [1.2, 1]$	2	0.00001
4.	Enzyme reaction, $x_0 = [0.25, 0.39, 0.415, 0.39]$	4	0.00001
5.	Stationary solution of a chemical reactor $x_0 = [1.09, 1.05, 3.05, 0.99, 6.05, 1.09]$	6	0.0001
6.	Robot kinematics problem $x_0 = [0.164, -0.98, -0.94, -0.32, -0.99, -0.05, 1.41, -0.91]$	8	0.0001
7.	Solar spectroscopy, $x_0 = [1, 1, 1, 1]$	4	0.001
8.	Estimation of parameters, $x_0 = [2.7, 90, 1500, 10]$	4	0.00001
9.	Propane combustion in air-reduced variant $x_0 = [10, 10, 0.05, 50.5, 0.05]$	5	0.00001
10.	Gear train with minimum inertia, $x_0 = [0.5, 0.5]$	2	0.00001
11.	Human heart dipole $x_0 = [0.299, 0.186, -0.0273, 0.0254, -0.474, 0.474, -0.0892, 0.0892]$	8	0.00001
12.	Neurophysiology, $x_0 = [0.01, 0.01, 0.01, 0.01, 0.01, 0.01]$	6	0.000001
13.	Combustion application, $x_0 = [1, 1, 1, 1, 1, 1, 1, 1, 1, 1]$	10	0.0000001
14.	Circuit design, $x_0 = [0.7, 0.5, 0.9, 1.9, 8.1, 8.1, 5.9, 1, 1.9]$	9	0.0001
15.	Thermistor, $x_0 = [0.01, 6100, 340]$	3	0.001
16.	Optimal design of a gear train, $x_0 = [15, 14, 35, 35]$	4	0.00001

and the accuracy for computing the solution (ε). The mathematical expression of these applications and the references, where these applications are described, are shown in Annex A. The rest of 124 problems are from different publications and their mathematical expressions are presented in Annex B. The number of variables of these problems is in the range [2, 50].

Tables 4.4 and 4.5 show the performances of DEEPS and of NELMEAD for solving this set of 140 problems, respectively.

Subject to the CPU time metric, we can see that DEEPS is 305 times faster than NELMED.

The algorithms compared in these numerical experiments find local solutions. Therefore, the comparisons are given in the following context. Let f_i^{ALG1} and f_i^{ALG2} be the minimal value found by ALG1 and ALG2 for the problems $i = 1,\ldots,140$, respectively. We say that, in the particular problem i the performance of ALG1 was better than the performance of ALG2 if:

$$\left| f_i^{ALG1} - f_i^{ALG2} \right| < 10^{-3} \qquad (4.1)$$

and if the number of iterations (#iter), or the number of function evaluations (#nfunc), or the CPU time of ALG1 was smaller than the number of iterations, or the number of function evaluations, or the CPU time corresponding to ALG2, respectively. The performances of the algorithms are displayed by the Dolan and

Table 4.4 Performance of DEEPS for solving 140 problems

A two-level random search method for unconstrained optimization

Total results for solving 140 problems

n	iter	nfunc	time	vfomin	vf0	fname
2	119	1485	0	−0.2644531378352E+03	−0.3747313737140E+02	1. WEBER-1
2	67	1318	0	0.9560744054913E+01	0.7859432489718E+02	2. WEBER-2
2	113	6198	1	0.8749847970082E+01	0.7860286479337E+02	3. WEBER-3
4	28	42903	2	0.3087657632221E−03	0.5313172272109E−02	4. ENZIMES
6	1375	6989983	197	0.7472925581554E−04	0.1961733675060E+08	5. REACTOR
8	43	2459	0	0.1828017526180E−07	0.5334258881257E+01	6. ROBOT
4	4	233	0	0.8316382216967E+01	0.9958700480657E+01	7. SPECTR
4	54	2705727	99	0.3185724691657E−01	0.2905300235663E+01	8. ESTIMP
5	10	10013013	274	0.2467602087429E−04	0.3312269269234E+08	9. PROPAN
2	15	342	0	0.1744152005590E+01	0.2563325000000E+04	10. GEAR-1
8	2404	36265585	1487	0.9966084682095E−04	0.1905692550902E+00	11. HHD
6	1044	7999853	219	0.8545018926146E−04	0.2391760016000E+02	12. NEURO
10	25	856	0	0.4061987800161E−08	0.1219988990749E+03	13. COMBUST
9	3731	9654682	1244	0.1036184837525E−03	0.2964578187893E+04	14. CIRCUIT
3	15	752179	251	0.1742216236340E+03	0.2335910048036E+10	15. THERM
4	6	3090	0	0.3886716443010E−13	0.7370818569964E−03	16. GEAR-2
2	13	1452	0	0.5223120133290E−08	0.2420000000000E+02	17. BANANA
2	19	19219	0	0.4898425368048E+02	0.4005000000000E+03	18. FRE-ROTH
2	8	409182	1	0.9096890029946E−10	0.1228198400000E+02	19. WHI-HOL
4	11	5546	1	0.5097821607941E−08	0.1300897728708E+02	20. MI-CAN
2	17	34183	1	0.3509702847612E−07	0.2820895744000E+04	21. HIMM-1
2	6	12055	0	−0.1031628453412E+01	0.3600768000000E+01	22. 3-CAMEL
2	13	1448	0	0.9837324056667E−10	0.6700000000000E+01	23. 6-CAMEL
4	41	210805	4	0.3415656241886E−07	0.1919200000000E+05	24. WOOD
2	73	39075	0	0.1100000000004E+02	0.9850440000000E+04	25. QUADR-2
2	25	126221	2	−0.1008600149530E+02	−0.4236176330972E+01	26. SHEKEL
8	29	1464284	119	0.9678327750962E−08	0.3180996976805E+02	27. DENSCHNA
2	13	112880	0	0.6599092624990E−11	0.6585000000000E+04	28. DENSCHNB
8	236	11823238	1036	0.2030922856474E−07	0.5220203819919E+04	29. DENSCHNC
2	3	151167	3	0.2505775587025E−09	0.7251294749507E+00	30. GRIEWANK
2	19	9774	0	0.6794175748481E−10	0.1990000000000E+03	31. BRENT
2	4	201408	2	0.2147910448310E−09	0.1700000000000E+02	32. BOOTH
2	3	151310	2	0.5443086223290E−10	0.1072000000000E+00	33. MATYAS
3	20	100171	2	0.1930369016075E−07	0.9615280400000E+06	34. COLVILLE
2	6	3002322	122	−0.9999999999144E+00	−0.3030892310248E−04	35. EASOM
8	181	21192	1	0.2962368797323E−04	0.3931547600000E+02	36. BEALE
4	5	2502034	50	0.7775209572113E−09	0.2150000000000E+03	37. POWELL
2	8	4010395	62	-0.1913222954963E+01	0.7090702573174E+01	38. McCORM
2	10	200879	2	0.4820864500816E−09	0.1060000000000E+03	39. HIMM-2
2	6	304418	2	0.1359029423164E−08	0.7490384000000E+03	40. LEON
2	4	101964	0	0.1946503233539E−11	0.3307193600000E+02	41. PRICE4
2	6	151312	2	−0.3791237214486E−02	0.2500000000000E+00	42. ZETTL
8	23	1171077	38	0.5509141299242E−08	0.6480000000000E+03	43. SPHERE
8	21	1069087	35	0.3867053460807E−08	0.2040000000000E+03	44. ELIPSOID
2	4	414840	2	0.5922563192900E+01	0.3330769000000E+07	45. HIMM-3

(continued)

Table 4.4 (continued)

n	iter	nfunc	time	vfomin	vf0	fname
3	3	302066	4	0.2120841536218E−08	0.8400000000000E+01	46. HIMM-4
2	5	10250	0	0.5247703287149E−10	0.4598493014643E+00	47. HIMM-5
2	6	3012421	15	−0.3523860737867E+00	0.8000000000000E+01	48. ZIRILLI
2	3	15519	0	−0.7833233140751E+02	−0.2000000000000E+02	49. STYBLIN
2	3	15480	0	−0.1999999999988E+01	0.7000000000000E+01	50. TRID
2	28	1404941	12	0.9942813112235E−08	0.8100000001600E+06	51. SCALQ
3	21	425993	6	0.8592871044861E−08	0.6290000000000E+03	52. SCHIT-1
6	25	508583	13	0.3292756425536E−07	0.7500000000000E+02	53. SCHIT-2
2	2	4070	0	0.7731990565053E+00	0.8768604814560E+02	54. SCHIT-3
5	4	4139	0	0.2541928088954E−08	0.3693847656250E+02	55. BROWN-A
4	6	60194	1	0.5918576884523E−09	0.4400000000000E+02	56. KELLEY
4	18	180402	4	0.2281055628062E−08	0.1890000000000E+03	57. NONSYS
2	2	10363	0	−0.1819999999993E+02	−0.1660000000000E+02	58. ZANGWILL
3	2	10392	1	−0.3923048451913E+00	0.1000000000000E+02	59. CIRCULAR
2	20	102703	3	0.4610244291330E−06	0.1082682265893E+01	60. POLEXP
2	1	5200	0	−0.5000000000000E+00	0.4000000000000E+00	61. DULCE
4	3	154209	12	0.5069703944580E−12	0.2266182511289E+01	62. CRAGLEVY
5	42	23205	1	0.3674886987132E−07	0.1600000000000E+02	63. BROYDEN1
8	40	2005756	84	0.9392975052918E+00	0.1900000000000E+02	64. BROYDEN2
50	126	654057	150	0.9960497839120E−05	0.6100000000000E+02	65. BROYDEN3
5	12	62010	1	0.6698885009208E−11	0.4856000000000E+04	66. FULRANK1
8	16	80264	5	0.2290198352507E−05	0.3846000000000E+05	67. FULRANK2
10	22	110320	8	0.3706326698194E−05	0.1076320000000E+06	68. FULRANK3
5	7	36160	5	0.2037153258108E−08	0.1165737899047E−01	69. TRIG-1
9	12	601804	154	0.1026965671220E−07	0.8201586330328E−01	70. TRIG-2
20	10	1001534	568	0.4625184409097E−07	0.3614762514618E+01	71. TRIG-3
10	10	25898	5	0.2098586473917E−07	0.1718986654839E+00	72. BROWN
4	20	203504	135	0.8582220162694E+05	0.7926693336997E+07	73. BRODEN
2	2	51661	1	−0.2345811576100E+01	−0.7664155024405E+00	74. HOSAKI
2	13	727	0	0.1544314623457E−09	0.6500000000000E+02	75. COSMIN
16	320	37518	3	0.4229792213475E+02	0.2712000000000E+04	76. BDQRTIC
10	55	28405	1	0.9099139103073E−01	0.1300000000000E+02	77. DIXON3DQ
4	56	283757	5	0.2495604370956E+01	0.1770000000000E+03	78. ENGVAL1
5	36	36596	1	0.1522038727647E+01	0.3011562500000E+04	79. EX-PEN
5	23	233749	6	0.5444996552378E+00	0.3700000000000E+02	80. BROYDENP
2	6	69034	3	0.9000000000179E+00	0.2402613308223E+01	81. TEO
2	6	30081	0	0.7415391850332E−02	0.1912500017166E+04	82. Coca
2	11	5676	0	−0.1249999999896E+01	0.5000000000000E+01	83. NEC
4	1	5212	0	0.8881666090321E−16	0.7324861284383E−02	84. QPEXP
8	10	51582	2	0.6822795617112E−10	0.4900000000000E+03	85. NONDQUAR
40	77	199656	33	0.1348798973067E−05	0.1170000000000E+03	86. ARWHEAD
4	1735	44990111	794	0.9879035371252E−04	0.1977237045971E+04	87. CUBE
5	641	3329422	71	0.6114213167963E−05	0.5800000000000E+03	88. NONSCOMP
10	56	3196	1	0.1765232882554E−05	0.2080000000000E+04	89. DENSCHNF
16	744	49811	4	0.1791687822047E−05	0.2000000000000E+01	90. BIGGSB1
30	14	72779	7	0.4367156132713E−06	0.2168583873406E+12	91. BORSEC6

(continued)

Table 4.4 (continued)

n	iter	nfunc	time	vfomin	vf0	fname
2	21	1926877	8	0.3686757105806E+02	0.1020162281649E+13	92. 3T-QUAD
3	18	1650269	12	0.1575758116274E−06	0.6976250929902E+27	93. MISHRA9
2	17	1559943	7	0.1775923522586E−08	0.6288740000000E+06	94. WAYBURN1
2	11	570063	2	0.1044643130973E−08	0.1079181850156E+03	95. WAYBURN2
10	64	331133	14	0.6666667207742E+00	0.1093660000000E+06	96. DIX&PRI
15	81	203113	13	0.9277471785896E−06	0.5200000000000E+03	97. QING
2	12	618989	3	−0.3873724182172E+04	0.4388316000000E+05	98. QUAD-2V
2	20	100338	1	-0.6850076846409E+02	0.4802237950000E+00	99. RUMP
4	60	311491	12	0.3995734044744E+00	0.2000450000000E+01	100. EX-CLIFF
10	83	42869	2	0.9898969787039E+00	0.3604000000000E+04	101. NONDIA
4	17	86897	3	−0.3499997998313E+01	0.2945148446828E+01	102. EG2
8	47	243103	8	0.1287394168882E−06	0.4680000000000E+04	103. LIARWHD
4	19	19373	2	0.1212871287130E+02	0.1660870312500E+07	104. FULL-HES
2	39	1954856	79	0.2807057782276E−10	0.1489929163060E−01	105. NALSYS
4	14	72067	1	−0.3739004994563E+02	0.2190000000000E+03	106. ENGVAL8
10	7	405	0	0.1000000012293E+01	0.7400000000000E+02	107. DIXMAANA
10	7	36106	10	0.1000000011413E+01	0.1380750000000E+03	108. DIXMAANB
5	6	30828	4	0.1000000001937E+01	0.1095000000000E+03	109. DIXMAANC
5	44	228394	4	0.7507319964675E−07	0.4005000000000E+04	110. DIAGAUP1
10	49	2811	1	-0.9499999615584E+01	0.2226137858737E+01	111. EG3-COS
10	66	2161	0	0.7730452422942E−07	0.2198553145800E+07	112. VARDIM
2	20	130469	1	−0.4999711763084E+00	0.3280000000000E+03	113. NARRCONE
10	14	16323	2	−0.2718122988146E+01	0.3875317250981E+01	114. ACKLEY
2	2	401736	5	−0.3885618082985E+01	0.8333333333333E+00	115. WOLFEmod
2	425	1087323	4	−0.6547921763872E+01	0.4161726395744E−03	116. PEAKS
3	79	800719	11	0.2492301853743E+02	0.1870000000000E+03	117. U18
2	1	1024	0	−0.1372874556465E−05	0.1082682265893E+01	118. U23
5	6	601032	16	0.6062571615241E−08	0.3000000000000E+02	119. SUMSQUAR
10	30	30467	2	0.1462958086757E−07	0.4830353687098E+13	120. VARDIM8
2	14	14192	1	0.7208138754557E−07	0.2706705664732E+00	121. MODULE
2	5	5076	0	0.8484175001490E−12	0.6064152299177E−01	122. PEXP
2	3	3041	0	0.6760954994400E−07	0.2706705817134E+00	123. COMB-EXP
2	31	31433	0	-0.5999999998409E+01	0.1990000000000E+04	124. QUADR1
2	8	4132	0	0.1000000000058E+02	0.6000000000000E+02	125. QUADR2
50	21	5397	3	0.4900000200180E+02	0.9841913092362E+03	126. SUM-EXP
2	679	1759498	15	−0.9552512351121E+04	0.2116666666667E+01	127. CAMEL
10	17	43784	1	0.2934784437013E−03	0.1100000000000E+02	128. SP-MOD
2	10	319	0	0.2024356506683E−09	0.1000000000000E+02	129. TRECANNI
10	23	2677	1	0.2020482388347E+03	0.2950000000000E+03	130. ProV-MV
10	16	1859	0	0.4233492307100E+02	0.5500000000000E+02	131. SCALQUAD
10	77	2468	0	0.6932289635176E+02	0.1224000000000E+04	132. BRASOV
10	8	20611	1	0.1389960254143E−05	0.5500000000000E+02	133. PROD-SUM
10	10	334	0	0.2180037938679E−05	0.1111111110100E+12	134. PRODPROD
10	74	189030	62	−0.6902392229727E+03	−0.4614893409920E+03	135. PS-COS
10	117	13636	5	−0.6617719095465E+02	−0.3773267891917E+01	136. PP-COS
10	122	13698	4	−0.6131657360927E+03	−0.2992116109892E+03	137. PS-SIN

(continued)

Table 4.4 (continued)

n	iter	nfunc	time	vfomin	vf0	fname
10	105	12237	4	−0.6929157853667E+02	−0.3515404061961E+01	138. PP-SIN
2	8	903	0	0.5817829773491E−08	0.3600000000013E+01	139. BOHA
2	129	14723	0	−0.2492930182626E+06	0.9997000016000E+04	140. DECK-AAR

Total number of iterations = 16911
Total number of function evaluations = 177973481
Total elapsed time (centiseconds) = 7706

Moré performance profiles (Dolan and Moré 2002). Figure 4.5 presents the performance profiles subject to the CPU computing time for solving 140 unconstrained optimization problems considered in this numerical study.

The percentage of problems for which an algorithm is the best is given on the left side of the plot. On the other hand, the right side of the plot gives the percentage of the problems that are successfully solved. In other words, for a given algorithm, the plot for $\tau = 1$ represents the fraction of problems, for which the algorithm was the most efficient over all algorithms. The plot for $\tau = \infty$ represents the fraction of problems solved by the algorithm irrespective of the required effort. Therefore, the plot for $\tau = 1$ is associated to the *efficiency* of the algorithm, while the plot for $\tau = \infty$ is associated to the *robustness* of the algorithm.

From Fig. 4.5, we can see that DEEPS is way more efficient and more robust than NELMED. Out of 140 problems, only for 106 of them does the criterion (4.1) hold. The table inside the plot shows the performances of the algorithms for $\tau = 1$, where #iter is the number of iterations for solving these problems, #nfunc is the number of evaluations of the function, and cpu is the CPU computing time. For example, subject to the CPU time metric, DEEPS was faster in solving 69 problems, that is, DEEPS achieved the minimum time in solving 69 problems out of 106, NELMED was faster in 18, and they obtained the same amount of time in solving 19 problems.

4.4   Performance of DEEPS for Solving Large-Scale Problems

In the following, let us present the performances of DEEPS for solving large-scale unconstrained optimization problems. We selected a number of 30 problems from our collection with different number of variables, but limited to 500. Table 4.6 presents the performances of DEEPS for solving these problems.

As we can see, DEEPS is able to solve large-scale unconstrained optimization problems, needing a reasonable number of iterations and of CPU time, but a huge number of evaluations of the minimizing function. This is typical for derivative-free methods.

Table 4.5 Performances of NELMED for solving 140 problems (FORTRAN77 version by O'Neill (1971), modified by John Burkardt)

n	iter	nfunc	time	vfomin	vf0	fname
2	4321	14260	1	-0.2644531398164E+03	-0.3747313737140E+02	1. WEBER-1
2	5749	17111	0	0.9560739834844E+01	0.7859432489718E+02	2. WEBER-2
2	485	1515	0	0.8749843722120E+01	0.7860286479337E+02	3. WEBER-3
4	58196	298649	8	0.3075923582382E−03	0.5313172272109E−02	4. ENZIMES
6	601448	2963265	50	0.3370929785002E−03	0.1961733675060E+08	5. REACTOR
8	18824	131208	2	0.6829467973914E−08	0.5334258881257E+01	6. ROBOT
4	44064	225895	35	0.6872367777921E+01	0.9958700480657E+01	7. SPECTR
4	220110632	999990007	26078	0.3822441651590E−01	0.2905300235663E+01	8. ESTIMP
5	16640	96881	1	0.2887516043798E−03	0.3312269269234E+08	9. PROPAN
2	3584	11541	1	0.1744152005589E+01	0.2563325000000E+04	10. GEAR-1
8	1232994	12451823	231	0.9315130266680E−05	0.1905692553768E+00	11. HHD
6	539699	4132548	52	0.2463694921725E−05	0.2391760016000E+02	12. NEURO
10	19060786	155723303	4155	0.8484442736605E−10	0.1219988990749E+03	13. COMBUST
9	39546	390639	71	0.1022362061183E−01	0.2964578187893E+04	14. CIRCUIT
3	54473	136162	83	0.1750977749414E+03	0.2335910048036E+10	15. THERM
4	8026086	36117397	533	0.7515500076120E−19	0.7370818569964E−03	16. GEAR-2
2	127791	414317	5	0.2734165291417E−08	0.2420000000000E+02	17. BANANA
2	7772	21242	0	0.4898425367977E+02	0.4005000000000E+03	18. FRE−ROTH
2	19981	66250	0	0.1646619899815E−09	0.1228198400000E+02	19. WHI-HOL
4	28366709	134400344	8779	0.2728619079879E−22	0.1300897728708E+02	20. MI-CAN
2	549	1801	0	0.2410661272377E−09	0.2820895744000E+04	21. HIMM-1
2	4138	12917	0	−0.1031628453487E+01	0.3600768000000E+01	22. 3-CAMEL
2	12681	40251	0	0.1791830653421E+01	0.6700000000000E+01	23. 6-CAME
4	9355302	47490787	408	0.9110670262491E−09	0.1919200000000E+05	24. WOOD
2	110941	357658	2	0.1100000000000E+02	0.9850440000000E+04	25. QUADR-2
2	13395	38020	0	0.5065439735221E+01	−0.4236176330972E+01	26. SHEKEL
8	32893597	239932420	16083	0.1913163675383E−15	0.3180996976805E+02	27. DENSCHNA
2	147386	476192	3	0.1325420598527E−13	0.6585000000000E+04	28. DENSCHNB
8	3098720	19502957	1747	0.1833616546785E+00	0.5220203819919E+04	29. DENSCHNC
2	711418	2298258	36	0.3219646771413E−14	0.7251294749507E+00	30. GRIEWANK
2	52622	184115	3	0.1293393700200E−14	0.1990000000000E+03	31. BRENT
2	52304	201225	1	0.4104965908510E−13	0.1700000000000E+02	32. BOOTH
2	717578	2004520	13	0.2799993833942E−14	0.1072000000000E+00	33. MATYAS
3	8195333	31914397	217	0.8289711921786E−07	0.9615280400000E+06	34. COLVILLE
2	1649974	5224920	131	−0.8110381996167E−04	−0.3030892310248E−04	35. EASOM
8	19783086	132373529	3919	0.6615259847323E−09	0.3931547600000E+02	36. BEALE
4	1559835	6876985	69	0.1285824459516E−08	0.2150000000000E+03	37. POWELL
2	4266	12315	0	−0.1913222954978E+01	0.7090702573174E+01	38. McCORM
2	340071	1067993	6	0.1445832251657E−14	0.1060000000000E+03	39. HIMM-2
2	63610	212134	2	0.1174796918677E−05	0.7490384000000E+03	40. LEON
2	1802097	6173224	39	0.1068218485126E−07	0.3307193600000E+02	41. PRICE4
2	78275	246644	1	−0.3791237220468E−02	0.2500000000000E+00	42. ZETTL
8	133777215	999990013	15717	0.1025689019328E+00	0.6480000000000E+03	43. SPHERE
8	26546059	200996384	3104	0.3701062404756E−13	0.2040000000000E+03	44. ELIPSOID
2	95	193	0	0.5924864255508E+01	0.3330769000000E+07	45. HIMM-3

(continued)

Table 4.5 (continued)

n	iter	nfunc	time	vfomin	vf0	fname
3	1439	4699	0	0.7544044265105E−06	0.8400000000000E+01	46. HIMM-4
2	1469	4880	0	0.4334193828917E−10	0.4598493014643E+00	47. HIMM-5
2	18886	63700	1	−0.1526394417735E+00	0.8000000000000E+01	48. ZIRILLI
2	285	769	0	−0.7833233140632E+02	−0.2000000000000E+02	49. STYBLIN
2	755782	2507508	17	−0.2000000000000E+01	0.7000000000000E+01	50. TRID
2	110285272	441135983	2383	0.1892243360928E−12	0.8100000001600E+06	51. SCALQ
3	7510	29043	1	0.1143027552724E−03	0.6290000000000E+03	52. SCHIT-1
6	966	4340	0	0.2273917223761E−08	0.7500000000000E+02	53. SCHIT-2
2	1029	3374	0	0.7731990567225E+00	0.8768604814560E+02	54. SCHIT-3
5	34744	190100	2	0.2297883703351E−06	0.3693847656250E+02	55. BROWN-A
4	81407052	368163139	3085	0.1422943777975E−13	0.4400000000000E+02	56. KELLEY
4	831618	3641092	35	0.1050959372381E−11	0.1890000000000E+03	57. NONSYS
2	8908	24481	1	−0.1820000000000E+02	-0.1660000000000E+02	58. ZANGWILL
3	3969	14757	0	−0.3923048452658E+00	0.1000000000000E+02	59. CIRCULAR
2	333330006	999990001	26315	0.3252892765785E−15	0.1082682265893E+01	60. POLEXP
2	18157983	63360019	360	−0.5000000000000E+00	0.4000000000000E+00	61. DULCE
4	95011309	435691620	26934	0.7500494352983E−17	0.2266182511289E+01	62. CRAGLEVY
5	732002	4006517	52	0.3752106106972E−14	0.1600000000000E+02	63. BROYDEN1
8	5888828	44813788	1273	0.2027502384709E−13	0.1900000000000E+02	64. BROYDEN2
50	28113950	999990028	129252	0.1254548102740E−03	0.6100000000000E+02	65. BROYDEN3
5	32105835	172990716	4058	0.5600659081408E−15	0.4856000000000E+04	66. FULRANK1
8	48480940	345462434	14258	0.1455367377274E−12	0.3846000000000E+05	67. FULRANK2
10	27953795	235714362	13299	0.2201500157965E−12	0.1076320000000E+06	68. FULRANK3
5	26934031	142029901	17864	0.2760740072636E−14	0.1165737899047E−01	69. TRIG-1
9	44504610	364293181	81627	0.4534850872176E−03	0.8201586330328E−01	70. TRIG-2
20	3717178	56764027	27604	0.6659567335342E−04	0.3614762514618E+01	71. TRIG-3
10	134984117	999990002	193939	0.3169785824748E−02	0.1718986654839E+00	72. BROWN
4	250	448	0	0.8582220220393E+05	0.7926693336997E+07	73. BRODEN
2	7205	21445	1	−0.1127794026972E+01	−0.7664155024405E+00	74. HOSAKI
2	19567	65820	0	0.7281603268129E−12	0.6500000000000E+02	75. COSMIN
16	6731934	88261512	4792	0.4229791375365E+02	0.2712000000000E+04	76. BDQRTIC
10	3233930	28345714	480	0.9090909101913E−01	0.1300000000000E+02	77. DIXON3DQ
4	67202284	307509056	4062	0.2495604366214E+01	0.1770000000000E+03	78. ENGVAL1
5	18287	107384	2	0.1522244423034E+01	0.3011562500000E+04	79. EX-PEN
5	44766	234527	6	0.3223477633005E−12	0.3700000000000E+02	80. BROYDENP
2	114222	324189	17	0.9000000000000E+00	0.2402613308223E+01	81. TEO
2	15239	44506	1	0.3603393492868E−11	0.1912500017166E+04	82. Coca
2	220243	770819	7	−0.1250000000000E+01	0.5000000000000E+01	83. NEC
4	249997502	999990008	21610	0.6498415264253E−22	0.7324861284383E−02	84. QPEXP
8	157576603	999990001	22699	0.1061987603079E+00	0.4900000000000E+03	85. NONDQUAR
40	45164548	999990036	100409	0.1740850637259E+00	0.1170000000000E+03	86. ARWHEAD
4	1221660	5840002	54	0.1170242033020E−03	0.1977237045971E+04	87. CUBE
5	164045317	883130211	11253	0.9253097139544E−09	0.5800000000000E+03	88. NONSCOMP
10	1255946	11901044	264	0.2432712919620E−11	0.2080000000000E+04	89. DENSCHNF
16	74285922	847283436	27520	0.4565388058408E−11	0.2000000000000E+01	90. BIGGSB1
30	60610021	999990003	86031	0.1288303836062E+03	0.2168583873406E+12	91. BORSEC6

(continued)

Table 4.5 (continued)

n	iter	nfunc	time	vfomin	vf0	fname
2	2424	7088	0	0.1273426289635E+03	0.1020162281649E+13	92. 3T-QUAD
3	148	300	0	0.5510270894397E−03	0.6976250929902E+27	93. MISHRA9
2	18567784	57590805	360	0.6454558509586E−14	0.6288740000000E+06	94. WAYBURN1
2	50268833	142015420	887	0.7079093388257E−15	0.1079181850156E+03	95. WAYBURN2
10	136976950	999990005	24547	0.1005563480381E+04	0.1093660000000E+06	96. DIX&PRI
15	3688046	45451698	1184	0.2261032544726E−09	0.5200000000000E+03	97. QING
2	1797	5885	0	−0.3873724182186E+04	0.4388316000000E+05	98. QUAD-2V
2	253	826	0	−0.1640345171681E+02	0.4802237950000E+00	99. RUMP
4	212439967	854777872	30989	0.3995732273681E+00	0.2000450000000E+01	100. EX-CLIFF
10	642708	4988085	93	0.5455027073350E−12	0.3604000000000E+04	101. NONDIA
4	3424718	16773818	726	−0.3447779105873E+01	0.2945148446828E+01	102. EG2
8	59110575	410824970	10345	0.1517431852535E−11	0.4680000000000E+04	103. LIARWHD
4	12770	57246	10	0.1212871287129E+02	0.1660870312500E+07	104. FULL-HES
2	1875451	5018254	183	0.1833800208498E−15	0.1489929163060E−01	105. NALSYS
4	104963	474395	4	−0.3739004995170E+02	0.2190000000000E+03	106. ENGVAL8
10	79384557	699911844	174704	0.1000000000000E+01	0.7400000000000E+02	107. DIXMAANA
10	115678296	999990003	252734	0.1001251911045E+01	0.1380750000011E+03	108. DIXMAANB
5	519054	2611878	315	0.1000000000000E+01	0.1095000000000E+03	109. DIXMAANC
5	207132	1046318	14	0.3533704907427E−12	0.4005000000000E+04	110. DIAGAUP1
10	685883	5148183	341	−0.8184677637534E+01	0.2226137858737E+01	111. EG3-COS
10	35829692	279603425	10440	0.4467510264809E−10	0.2198551162500E+07	112. VARDIM
2	3620863	12323786	77	−0.4999999976812E+00	0.3280000000000E+03	113. NARRCONE
10	281032	2414390	211	0.3841368041779E+01	0.3875317250981E+01	114. ACKLEY
2	22403	74302	0	−0.3885618083164E+01	0.8333333333333E+00	115. WOLFEmod
2	2316670	7627830	281	−0.4105766136015E−05	0.4161726395744E−03	116. PEAKS
3	117079	492835	3	0.2492301853327E+02	0.1870000000000E+03	117. U18
2	13116974	39350868	846	0.3541982800515-316	0.1082682265893E+01	118. U23
5	168469	738431	10	0.6048706368964E−13	0.3000000000000E+02	119. SUMSQUAR
10	17305874	131238244	3066	0.3070835264168E−10	0.4830353687098E+13	120. VARDIM8
2	74633097	223899102	5848	0.2285955775885-316	0.2706705664732E+00	121. MODULE
2	69223899	207671069	4134	0.8913098597084-315	0.6064152299177E−01	122. PEXP
2	140380939	491473848	33963	0.8783252018943-318	0.2706705817134E+00	123. COMB-EXP
2	47128	153693	1	−0.6000000000000E+01	0.1990000000000E+04	124. QUADR1
2	311846	992749	6	0.1000000000000E+02	0.6000000000000E+02	125. QUADR2
50	9331609	334998481	128011	0.4900000000000E+02	0.9841913092362E+03	126. SUM-EXP
2	874	1606	0	−0.9552512351121E+04	0.2116666666667E+01	127. CAMEL
10	70892670	898652510	48230	0.6043245601822E−06	0.1100000000000E+02	128. SP-MOD
2	68676	210584	1	0.1519137994603E−13	0.1000000000000E+02	129. TRECANNI
10	26005289	220984101	6272	0.1925000000000E+03	0.2950000000000E+03	130. ProV-MV
10	321864	2377527	83	0.4125000000000E+02	0.5500000000000E+02	131. SCALQUAD
10	25695259	222840459	7139	0.6932289628180E+02	0.1224000000000E+04	132. BRASOV
10	64445457	728720902	39364	0.3730283454221E−07	0.5500000000000E+02	133. PROD-SUM
10	351348016	999990005	90615	0.1642785421782E+05	0.1111111110100E+12	134. PRODPROD
10	10055497	93021910	28241	−0.5281026787308E+03	−0.4614893409920E+03	135. PS-COS
10	74217185	689414382	225414	0.5992685311919E+02	−0.3773267891917E+01	136. PP-COS
10	10313102	132740842	42701	−0.6131657364757E+03	−0.2992116109892E+03	137. PS-SIN

(continued)

Table 4.5   (continued)

n	iter	nfunc	time	vfomin	vf0	fname
10	92307224	846888382	279080	-0.5054277692864E+02	-0.3515404061961E+01	138. PP-SIN
2	29579	96784	1	0.2287500351344E+01	0.3600000000013E+01	139. BOHA
2	1931	6434	0	-0.2492930182630E+06	0.9997000016000E+04	140. DECK-AAR

Total CPU time = 2354628 centiseconds ≈ 6.54 h

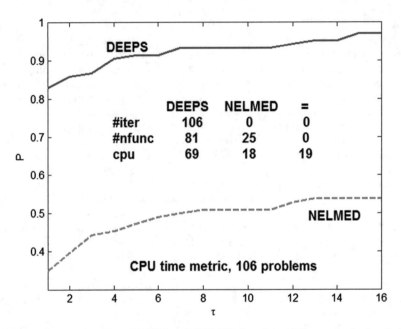

Fig. 4.5 Performance profiles of DEEPS and NELMED for solving 140 unconstrained optimization problems

Table 4.6 Performances of DEEPS for solving 30 large-scale problems

A two-level random search method for unconstrained optimization

Total results for solving 30 large-scale problems

n	iter	nfunc	time	vfomin	vf0	fname
100	54	2726250	2813	0.1933041622350E−05	0.3976246221006E+03	27. DENSCHNA
200	475	4178799	3365	0.7607056471717E−05	0.7129000000000E+04	28. DENSCHNB
200	39	1977715	1582	0.1906786878984E−05	0.4027200000000E+01	33. MATYAS
100	594	3025219	1352	0.8927972405540E−06	0.3383500000000E+06	44. ELIPSOID
200	1184	700626	1069	0.1869593904842E−03	0.2110000000000E+03	63. BROYDEN1
200	10	52095	399	0.3927344805542E−05	0.6972774389540E+04	69. TRIG-1
200	14	36127	296	0.9136497330827E−05	0.3800870492365E+01	72. BROWN
200	498	28611	38	0.6234069231056E−05	0.6500000000000E+04	75. COSMIN
200	56	285028	245	0.2460315490512E+03	0.1174100000000E+05	78. ENGVAL1
500	380	386099	810	0.4266196551499E+03	0.1746550388459E+16	79. EX-PEN
500	1179	12010364	41069	0.7896764624658E−03	0.5320000000000E+03	80. BROYDENP
500	13	67719	90	0.3947652312005E−08	0.4034200000000E+05	85. NONDQUAR
400	508	1316673	2942	0.1520943530238E−03	0.1197000000000E+04	86. ARWHEAD
500	2738	156891	617	0.4679348483033E−02	0.1040000000000E+06	89. DENSCHNF
500	17	87802	268	0.1515258658432E−03	0.8274826754904E+26	91. BORSEC6
100	687	3499045	1676	0.2427940007984E−04	0.5850000000000E+05	103. LIARWHD
400	41	41802	508	0.1212871290753E+02	0.6936927728460E+14	104. FULL-HES
500	28	1598	24	0.1000032590334E+01	0.4063500000000E+04	107. DIXMAANA
500	10	51691	1003	0.1000030776510E+01	0.7495937500000E+04	108. DIXMAANB
500	22	113066	2723	0.1000029993149E+01	0.1304550000000E+05	109. DIXMAANC
100	241	1251550	524	0.2597022323503E−04	0.8010000000000E+05	110. DIAGAUP1
500	3332	190828	821	−0.4994993713630E+03	0.1412206087357E+03	111. EG3-COS
100	301	1533231	658	0.1333805427545E−04	0.1310583696893E+15	112. VARDIM
500	50	5006609	18461	0.7079248254927E−02	0.2505000000000E+06	119. SUMSQUAR
100	238	1212244	523	0.5671097610006E−05	0.1717629476525E+29	120. VARDIM8
500	28	7193	40	0.4990001347310E+03	0.1002268292467E+05	126. SUM-EXP
500	22	56795	109	0.9444412279940E−01	0.5010000000000E+03	128. SP-MOD
100	5001	550642	222	0.1990001436291E+06	0.3284500000000E+06	130. ProV-MV
200	3116	99435	91	0.1589333398727E+04	0.2706400000000E+05	132. BRASOV
300	16	41334	660	0.1519490671575E−02	0.4515000000000E+05	133. PROD-SUM

Total number of iterations = 20892

Total number of function evaluations = 40693081

Total elapsed time (centiseconds) = 84998

References

Abramson, M.A., (2005). Second-order behavior of pattern search. SIAM Journal on Optimization, 16(2), 315-330.

Bongartz, I., Conn, A.R., Gould, N.I.M., & Toint, Ph.L, (1995). CUTE: constrained and unconstrained testing environments. ACM Transactions on Mathematical Software, 21, 123-160.

Dolan, E.D., & Moré, J.J., (2002). Benchmarking optimization software with performance profiles. Mathematical Programming, 91, 201-213.

Frimannslund, L., & Steihaug, T., (2007). A generating set search method using curvature information. Computational Optimization and Applications, 38, 105-121.

Frimannslund, L., & Steihaug, T., (2011). On a new method for derivative free optimization, International Journal on Advances in Software, 4(3-4), 244-255.

Higham, N. (n.d.) The matrix Computation Toolbox. http://www.ma.man.ac.uk/~higham/mctoolbox/

O'Neill, R., (1971). Algorithm AS 47: Function Minimization Using a Simplex Procedure. Applied Statistics, Volume 20(3), pp. 338-345.

Powell, M.J.D., (2004). *The NEWUOA software for unconstrained optimization without derivatives.* Department of Applied Mathematics and Theoretical Physics, University of Cambridge. DAMTP 2004/NA05, November.

Wolfe, P., (1971). Convergence conditions for ascent methods. II: Some corrections. SIAM Review, 13(2), 185-188.

Chapter 5
Conclusion

For solving unconstrained optimization problems, a new derivative-free algorithm is presented. The idea of this algorithm is to capture a deep view of the landscape of the minimizing function around the initial point by randomly generating some trial points at two levels. This is a new approach that has not been covered up till now. Basically, our algorithm is a pure random search, but at two levels.

The algorithm is based on the selection of the minimum point of the minimizing function from a set of N trial points randomly generated in a domain D specified by some bounds *lobnd* and *upbnd*, the same for all variables. This set of trial points is called *complex*. The cardinal of the complex may be different from the number of variables of the minimizing function. The algorithm performs a deep exploration of the domain D at two levels. Around each of the trail points $x_j \in D \subset \mathbb{R}^n, j = 1, \ldots, N$, from the first level, another set of M local trial points from the second level is generated in certain local domains $D_j \subset \mathbb{R}^n, j = 1, \ldots, M$, specified by some bounds *lobndc* and *upbndc*, the same for all domains and variables. Usually, the bounds of the local domains are smaller than the bounds of the domain D. The trial points are replaced by points from the local domains, for which the value of the minimizing function is smaller. Selection of the best values for N and M is an open problem. However, our intensive numerical experiments showed that the best results are obtained when N and M are smaller or equal to n and $N \leq M$. The bounds defining the domain D and the local domains $D_j, j = 1, \ldots, M$, have a small influence on the performances of the algorithm. For a given optimization problem, the best values for these bounds is also an open question. The best results are obtained for reasonably small domains around the initial point, emphasizing once more the local character of the algorithm.

At every iteration, from all the trial and the local trial points, the algorithm selects the point corresponding to the minimum value of the minimizing function. The set of trail points is modified by some specific procedures often used in derivative-free

© The Author(s), under exclusive license to Springer Nature Switzerland AG 2021
N. Andrei, *A Derivative-free Two Level Random Search Method for Unconstrained Optimization*, SpringerBriefs in Optimization,
https://doi.org/10.1007/978-3-030-68517-1_5

optimization methods, which are based on: computing the middle points and the reflection of points, reducing the bounds of the domain D or of the local domains associated to each trial points, reducing the complex or on the line searches. Obviously, the above procedures have been used for modifying the trial points in our algorithm, but some other procedures may also be proposed and used. The idea is to apply those procedures for modifying the trial points, which keep the number of function evaluations as small as possible.

At some specific iterations, the bounds of the domain D and the bounds of the local domains $D_j, j = 1, \ldots, M$, are reduced by their halving. It has been proved that along the iterations, the maximum distance among the trial points tends to zero, and the maximum distance among the trial points and the local trial points also tends to zero, these ensuring the convergence of the algorithm.

The algorithm has two phases: the reducing phase and the stalling phase. In the reducing phase, for a relatively small number of iterations, the algorithm produces an abrupt reduction of the function values. In this phase, there are some iterations defining a plateau, where the reduction of the function values is not significant. These plateaus start from the iterations, where the bounds of the domains, in which the trial points are generated, are reduced (halved), or where the complex is reduced. The stalling phase is characterized by a large number of iterations, where the function values change very slowly, converging to its minimum value. It is worth mentioning that the behavior of our algorithm is very close to the steepest descent algorithm for unconstrained optimization. In the steepest descent algorithm, there is also a reduction phase, where the function values are drastically reduced, followed by a stalling phase, where, for a relatively large number of iterations, the function values change slowly. The main difference between our algorithm and the steepest descent is that in the reduction phase of our algorithm, the reduction of the function values presents some plateaus, where the function values are very little reduced.

All in all, the algorithm produces a sequence of points $\{x_k\}$, for which $f(x_{k+1}) \leq f(x_k)$ at any iteration $k = 1, 2, \ldots$. Since there is no connection among the points randomly generated along the iterations, it is not possible to state any convergence property of the algorithm. However, supposing that the function is continuous and bounded below, having in view that for any $k = 1, 2, \ldots, f(x_{k+1}) \leq f(x_k)$, it follows that the corresponding sequence $\{x_k\}$ is convergent to a point x^*, for which $f(x^*) \leq f(x_0)$. Nothing can be said about the optimality of x^*, but this can be said about all derivative-free optimization algorithms.

In this form, the algorithm has a number of open problems that influence its performances. These refer to the selection of: the number of the trial points, the number of the local trial points, the values of the bounds of the domain, where the trial points are generated, and the values of the bounds of the local domains, where the local trial points are generated. All these need further investigations.

Intensive numerical experiments have shown that the suggested algorithm works very well in practice, being able to solve a large variety of unconstrained optimization problems up to 500 variables, as well as real applications from different areas of activity with a small number of variables. Numerical comparisons versus the algo-rithm of Nelder-Mead in implementation of O'Neill have proved its superiority subject to the number of function evaluations and to the CPU computing time. Comparisons of DEEPS versus some other random derivative-free optimization algorithms follow to be investigated.

Annex A

List of Applications

The mathematical expression of the applications from Table 4.3 solved by DEEPS and NELMED are as follows:

(1) *Weber (1)* (Kelley 1999, pp. 118–119), (Andrei 2003, p. 58) [WEBER-1]

$$f(x) = 2\sqrt{(x_1 - 2)^2 + (x_2 - 42)^2} + 4\sqrt{(x_1 - 90)^2 + (x_2 - 11)^2}$$
$$+ 5\sqrt{(x_1 - 43)^2 + (x_2 - 88)^2}.$$

Initial point: $x_0 = [10, \ 10]$. $f(x_0) = -0.374731\text{E} + 02$.

(2) *Weber (2)* (Kelley 1999, pp. 118–119) [WEBER-2]

$$f(x) = 2\sqrt{(x_1 + 10)^2 + (x_2 + 10)^2} - 4\sqrt{(x_1)^2 + (x_2)^2}$$
$$+ 2\sqrt{(x_1 - 5)^2 + (x_2 - 8)^2} + \sqrt{(x_1 - 25)^2 + (x_2 - 30)^2}.$$

Initial point: $x_0 = [1.2, \ 1]$. $f(x_0) = 0.785943\text{E} + 02$.

(3) *Weber (3)* (Kelley 1999, pp. 118–119) [WEBER-3]

$$f(x) = 2\sqrt{(x_1 + 10)^2 + (x_2 + 10)^2} - 4\sqrt{(x_1)^2 + (x_2)^2}$$
$$+ 2\sqrt{(x_1 - 5)^2 + (x_2 - 8)^2} + \sqrt{(x_1 - 25)^2 + (x_2 - 30)^2}$$
$$+ \sin\left(0.0035(x_1^2 + x_2^2)\right).$$

Initial point: $x_0 = [1.2, \ 1]$. $f(x_0) = 0.786028\text{E} + 02$.

© The Author(s), under exclusive license to Springer Nature Switzerland AG 2021
N. Andrei, *A Derivative-free Two Level Random Search Method for Unconstrained Optimization*, SpringerBriefs in Optimization,
https://doi.org/10.1007/978-3-030-68517-1

(4) *Analysis of Enzymes Reaction* (Andrei 2003, p. 62) [ENZIMES]

$$f(x) = \sum_{i=1}^{11} \left(y_i - \frac{x_1 \left(u_i^2 + u_i x_2 \right)}{u_i^2 + u_i x_3 + x_4} \right)^2,$$

where y_i and u_i have the following values:

i	y_i	u_i	i	y_i	u_i
1	0.1957	4.000	7	0.0456	0.125
2	0.1947	2.000	8	0.0342	0.100
3	0.1735	1.000	9	0.0323	0.0833
4	0.1600	0.500	10	0.0235	0.0714
5	0.0844	0.250	11	0.0246	0.0625
6	0.0627	0.167			

Initial point: $x_0 = [0.25, 0.39, 0.415, 0.39]$. $f(x_0) = 0.531317E - 02$.

(5) *Stationary Solution of a Chemical Reactor* (Shacham 1986, pp. 1455–1481) [REACTOR]

$$\begin{aligned}
f(x) = &(1 - x_1 - k_1 x_1 x_6 + r_1 x_4)^2 \\
&+ (1 - x_2 - k_2 x_2 x_6 + r_2 x_5)^2 \\
&+ (-x_3 + 2k_3 x_4 x_5)^2 \\
&+ (k_1 x_1 x_6 - r_1 x_4 - k_3 x_4 x_5)^2 \\
&+ (1, 5(k_2 x_2 x_6 - r_2 x_5) - k_3 x_4 x_5)^2 \\
&+ (1 - x_4 - x_5 - x_6)^2,
\end{aligned}$$

where, $k_1 = 31.24$, $k_2 = 0.272$, $k_3 = 303.03$, $r_1 = 2.062$, $r_2 = 0.02$.

Initial point: $x_0 = [1.09, 1.05, 3.05, 0.99, 6.05, 1.09]$. $f(x_0) = 0.196173E + 08$.

(6) *Robot Kinematics Problem* (Kearfott 1990, pp. 152–157), (Andrei 2013, pp. 101–103), (Floudas, Pardalos, et al. 1999, pp. 329–331) [ROBOT]

$$\begin{aligned}
f(x) = &\left(4.731.10^{-3} x_1 x_3 - 0.3578 x_2 x_3 - \right. \\
&\left. 0.1238 x_1 x_7 - 1.637.10 x_2 - 0.6734 x_4 - 0.6022\right)^2 \\
&+ (0.2238 x_1 x_3 + 0.7623 x_2 x_3 + \\
&\quad 0.2638 x_1 - x_7 - 0.07745 x_2 - 0.6734 x_4 - 0.6022)^2 \\
&+ \left(x_6 x_8 + 0.3578 x_1 + 4.731.10^{-3} x_2\right)^2 \\
&+ (-0.7623 x_1 + 0.2238 x_2 + 0.3461)^2 \\
&+ \left(x_1^2 + x_2^2 - 1\right)^2 \\
&+ \left(x_3^2 + x_4^2 - 1\right)^2 \\
&+ \left(x_5^2 + x_6^2 - 1\right)^2 \\
&+ \left(x_7^2 + x_8^2 - 1\right)^2.
\end{aligned}$$

Initial point: $x_0 = [0.164, -0.98, -0.94, -0.32, -0.99, -0.05, 0.41, -0.91]$. $f(x_0) = 0.533425E + 01$.

(7) *Solar Spectroscopy* (Andrei 2003, p. 68) [SPECTR]

$$f(x) = \sum_{i=1}^{13} \left(x_1 + x_2 \exp\left(-\frac{(i + x_3)^2}{x_4} \right) - y_i \right)^2,$$

where y_i, $i = 1, \ldots, 13$ are as in the below table:

i	y_i	i	y_i
1	0.5	8	2.5
2	0.8	9	1.6
3	1	10	1.3
4	1.4	11	0.7
5	2	12	0.4
6	2.4	13	0.3
7	2.7		

Initial point: $x_0 = [1, 1, 1, 1]$. $f(x_0) = 0.995870E + 01$.

(8) *Estimation of Parameters* (Himmelblau 1972, p. 430]) [ESTIMP]

$$f(x) = \sum_{i=1}^{7} \left(\frac{x_1^2 + a_i x_2^2 + a_i^2 x_3^2}{(1 + a_i x_4^2) b_i} - 1 \right)^2,$$

where the parameters a_i, b_i, $i = 1, \ldots, 7$ have the following values:

i	a_i	b_i
1	0.0	7.391
2	0.000428	11.18
3	0.0010	16.44
4	0.00161	16.20
5	0.00209	22.20
6	0.00348	24.02
7	0.00525	31.32

Initial point: $x_0 = [2.7, 90, 1500, 10]$. $f(x_0) = 0.290530E + 01$.

(9) *Propane Combustion in Air-Reduced Variant* (Meintjes and Morgan 1990, pp. 143–151; Averick et al. 1992, pp.18–19; Andrei 2013, pp. 54–56; Floudas, Pardalos et al. 1999, p. 327) [PROPAN]

$$f(x) = (x_1x_2 + x_1 - 3x_5)^2$$
$$+ \left(2x_1x_2 + x_1 + 2R_{10}x_2^2 + x_2x_3^2 + R_7x_2x_3 + R_9x_2x_4 + R_8x_2 - Rx_5\right)^2$$
$$+ \left(2x_2x_3^2 + R_7x_2x_3 + 2R_5x_3^2 + R_6x_3 - 8x_5\right)^2$$
$$+ \left(R_9x_2x_4 + 2x_4^2 - 4Rx_5\right)^2$$
$$+ \left(x_1x_2 + x_1 + R_{10}x_2^2 + x_2x_3^2 + R_7x_2x_3 + R_9x_2x_4 + R_8x_2 + R_5x_3^2 + R_6x_3 + x_4^2 - 1\right)^2,$$

where

$$R_5 = 0.193 \qquad\qquad R_6 = 0.4106217541E - 3$$
$$R_7 = 0.5451766686E - 3$$
$$R_8 = 0.44975E - 6 \qquad R_9 = 0.3407354178E - 4$$
$$R_{10} = 0.9615E - 6$$
$$R = 10$$

Initial point: $x_0 = [10,\ 10,\ 0.05,\ 50.5,\ 0.05]$. $f(x_0) = 0.331226E + 08$.

(10) *Gear Train of Minimum Inertia* (Sandgren and Ragsdell 1980; Schittkowski 1987, Problem 328, p. 149) [GEAR-1]

$$f(x) = 0.1\left(12 + x_1^2 + \left(1 + x_2^2\right)/x_1^2 + \left(x_1^2x_2^2 + 100\right)/x_1^4x_2^4\right).$$

Initial point: $x_0 = [0.5,\ 0.5]$. $f(x_0) = 0.256332E + 04$.

(11) *Human Heart Dipole* (Andrei 2003, p. 65; Andrei 2013, pp. 51–54; Averick et al. 1992, p. 17; Nelson and Hodgkin 1981, pp. 817–823) [HHD]

$$f(x) = \ (x_1 + x_2 - s_{mx})^2$$
$$+ \left(x_3 + x_4 - s_{my}\right)^2$$
$$+ \left(x_1x_5 + x_2x_6 - x_3x_7 - x_4x_8 - s_A\right)^2$$
$$+ \left(x_1x_7 + x_2x_8 + x_3x_5 + x_4x_6 - s_B\right)^2$$
$$+ \left(x_1\left(x_5^2 - x_7^2\right) - 2x_3x_5x_7 + x_2\left(x_6^2 - x_8^2\right) - 2x_4x_6x_8 - s_C\right)^2$$
$$+ \left(x_3\left(x_5^2 - x_7^2\right) + 2x_1x_5x_7 + x_4\left(x_6^2 - x_8^2\right) + 2x_2x_6x_8 - s_D\right)^2$$
$$+ \left(x_1x_5\left(x_5^2 - 3x_7^2\right) + x_3x_7\left(x_7^2 - 3x_5^2\right) + x_2x_6\left(x_6^2 - 3x_8^2\right) + x_4x_8\left(x_8^2 - 3x_6^2\right) - s_E\right)^2$$
$$+ \left(x_3x_5\left(x_5^2 - 3x_7^2\right) - x_1x_7\left(x_7^2 - 3x_5^2\right) + x_4x_6\left(x_6^2 - 3x_8^2\right) - x_2x_8\left(x_8^2 - 3x_6^2\right) - s_F\right)^2,$$

where

$s_{mx} = 0.485$	$s_A = -0.0581$	$s_C = 0.105$	$s_E = 0.167$
$s_{my} = -0.0019$	$s_B = 0.015$	$s_D = 0.0406$	$s_F = -0.399$.

Initial point: $x_0 = [0.299 \quad 0.186 \quad -0.0273 \quad 0.0254 \quad -0.474 \quad 0.474 \quad -0.0892$
$0.0892]^T$

$$f(x_0) = 0.190569E + 00.$$

(12) *Neurophysiology* (Andrei 2013, pp. 57–61; Verschelde et al. 1994, pp. 915–930) [NEURO]

$$f(x) = \left(x_1^2 + x_3^2 - 1\right)^2 + \left(x_2^2 + x_4^2 - 1\right)^2$$
$$+ \left(x_5 x_3^3 + x_6 x_4^3 - 1\right)^2 + \left(x_5 x_1^3 + x_6 x_2^3 - 2\right)^2$$
$$+ \left(x_5 x_1 x_3^2 + x_6 x_2 x_4^2 - 1\right)^2 + \left(x_5 x_3 x_1^2 + x_6 x_4 x_2^2 - 4\right)^2.$$

Initial point: $x_0 = [0.01, \ldots, 0.001]$. $\quad f(x_0) = 0.239991E + 02$.

(13) *Combustion Application* (Morgan 1987; Andrei 2013, pp. 61–63) [COMBUST]

$$f(x) = \left(x_2 + 2x_6 + x_9 + 2x_{10} - 10^{-5}\right)^2$$
$$+ \left(x_3 + x_8 - 3 \cdot 10^{-5}\right)^2$$
$$+ \left(x_1 + x_3 + 2x_5 + 2x_8 + x_9 + x_{10} - 5 \cdot 10^{-5}\right)^2$$
$$+ \left(x_4 + 2x_7 - 10^{-5}\right)^2$$
$$+ \left(0.5140437 \cdot 10^{-7} x_5 - x_1^2\right)^2$$
$$+ \left(0.1006932 \cdot 10^{-6} x_6 - 2x_2^2\right)^2$$
$$+ \left(0.7816278 \cdot 10^{-15} x_7 - x_4^2\right)^2$$
$$+ \left(0.1496236 \cdot 10^{-6} x_8 - x_1 x_3\right)^2$$
$$+ \left(0.6194411 \cdot 10^{-7} x_9 - x_1 x_2\right)^2$$
$$+ \left(0.2089296 \cdot 10^{-14} x_{10} - x_1 x_2^2\right)^2.$$

Initial point: $x_0 = [1, 1, 1, 1, 1, 1, 1, 1, 1, 1]$. $\quad f(x_0) = 0.017433088$.

(14) *Circuit Design* (Ratschek and Rokne 1993, p. 501; Andrei 2009, pp. 243–244; Price 1978, pp. 367–370) [CIRCUIT]

$$f(x) = (x_1 x_3 - x_2 x_4)^2 + \sum_{k=1}^{4} \left(a_k^2 + b_k^2\right),$$

where

$$a_k = (1 - x_1x_2)x_3\{\exp\left[x_5\left(g_{1k} - g_{3k}x_7 \cdot 10^{-3} - g_{5k}x_8 \cdot 10^{-3}\right)\right] - 1\}$$
$$+ g_{4k}x_2 - g_{5k}, \quad k = 1, \ldots, 4,$$

$$b_k = (1 - x_1x_2)x_4\{\exp\left[x_6\left(g_{1k} - g_{2k} - g_{3k}x_7 \cdot 10^{-3} - g_{4k}x_9 \cdot 10^{-3}\right)\right] - 1\}$$
$$+ g_{4k} - g_{5k}x_1, \quad k = 1, \ldots, 4,$$

$$g = \begin{bmatrix} 0.4850 & 0.7520 & 0.8690 & 0.9820 \\ 0.3690 & 1.2540 & 0.7030 & 1.4550 \\ 5.2095 & 10.0677 & 22.9274 & 20.2153 \\ 23.3037 & 101.7790 & 111.4610 & 191.2670 \\ 28.5132 & 111.8467 & 134.3884 & 211.4823 \end{bmatrix},$$

Initial   point: x_0 =   [0.7, 0.5, 0.9, 1.9, 8.1, 8.1, 5.9, 1, 1.9].
$f(x_0) = 0.296457\mathrm{E} + 04.$

(15) *Thermistor* (Andrei 2009, pp. 722–723) [THERM]

$$f(x) = \sum_{i=1}^{16} \left(y_i - x_1 \exp\left(\frac{x_2}{45 + 5i + x_3}\right)\right)^2,$$

where

i	y_i	i	y_i
1	34780	9	8261
2	28610	10	7030
3	23650	11	6005
4	19630	12	5147
5	16370	13	4427
6	13720	14	3820
7	11540	15	3307
8	9744	16	2872

Initial point: $x_0 = [0.01, 6100, 340].$ $f(x_0) = 0.233591\mathrm{E} + 10.$

(16) *Optimal Design of a Gear Train* (Sandgren 1988, pp. 95–105; Andrei 2013, p. 79) [GEAR-2]

$$f(x) = \left(\frac{1}{6.931} - \frac{x_1x_2}{x_3x_4}\right)^2.$$

Initial point: $x_0 = [15, 14, 35, 35].$ $f(x_0) = 0.737081\mathrm{E} - 03.$

References

Andrei, N., (2003). *Modele, Probleme de Test şi Aplicaţii de Programare Matematică.* [*Models, Test problems and Applications of Mathematical Programming*], Editura Tehnică [Technical Press], Bucharest, Romania.

Andrei, N., (2009). *Critica Raţiunii Algoritmilor de Optimizare fără Restricţii.* [*Criticism of the Unconstrained Optimization Algorithms Reasoning*], Editura Academiei Române [Academy Publishing House], Bucharest, Romania.

Andrei, N., (2013). *Nonlinear Optimization Applications using the GAMS Technology.* Springer Science + Business Media, New York.

Averick, B.M., Carter, R.G., Moré, J.J., & Xue, G.L., (1992). *The MINPACK-2 test problem collection.* Mathematics and Computer Science Division, Argonne National Laboratory, Preprint MCS-P153-0692. Argonne, USA.

Floudas, C.A., Pardalos, M.P., Adjiman, C.S., Esposito, W.R., Gümüs, Z.H., Harding, S.T., Klepeis, J.L., Meyer, C.A., & Schweiger, C.A., (1999). *Handbook of Test Problems in Local and Global Optimization*, Kluwer Academic Publishers, Dordrecht.

Himmelblau, D.M., (1972). *Applied Nonlinear Programming.* McGraw-Hill, New York.

Kearfott, R., (1990). Novoa, M.: INTBIS, a portable interval Newton bisection package. ACM Trans. Math. Software, 16, 152-157.

Kelley, C.T., (1999). *Iterative Methods for Optimization.* SIAM, Frontiers in Applied Mathematics, Philadelphia, USA.

Meintjes, K., & Morgan, A.P., (1990). Chemical-equilibrium systems as numerical test problems. ACM Trans. Math. Software, 16, 143-151.

Morgan, A.P. (1987). *Solving Polynomial Systems Using Continuation for Scientific and Engineering Problems.* Englewood Cliff: Prentice Hall.

Nelson, C.V., & Hodgkin, B.C., (1981). Determination of magnitudes, directions and locations of two independent dipoles in a circular conducting region from boundary potential measurements. IEEE Transactions on Biomedical Engineering, 28, 817-823.

Price, W.L., (1978). A controlled random search procedure for global optimization. The Computer Journal, 20(4), 367-370.

Ratschek, H., & Rokne, J., (1993). A circuit design problem. J. Global Optimization, 3, 501.

Sandgren, E., (1988). Nonlinear integer and discrete programming in mechanical design. In: Proceedings of the ASME Design Technology Conference, Kissimme, 95-105.

Sandgren, E., & Ragsdell, K.M., (1980). The utility of nonlinear programming algorithms: A comparative study – Part I. Journal of Mechanical Design, 102(3), 540-546.

Schittkowski, K., (1987). *More Test Examples for Nonlinear Programming Codes.* Springer Verlag, Berlin.

Shacham, M., (1986). Numerical solution of constrained nonlinear algebraic equations. Int. Journal for Numerical Methods in Engineering, 23, 1455-1481.

Verschelde, J., Verlinden, P., & Cools, R., (1994). Homotopies exploiting Newton polytopes for solving sparse polynomial systems. SIAM Journal on Numerical Analysis, 31, 915-930.

Annex B

List of Test Functions

The mathematical expression of the test functions considered in our numerical experiments is as follows:

(17) *Rosenbrock: Valley of Banana Function* (Andrei 2003, p. 39) [BANANA]

$$f(x) = \sum_{i=1}^{n/2} c\left(x_{2i} - x_{2i-1}^2\right)^2 + \left(1 - x_{2i-1}\right)^2, \quad c = 1000$$

$n = 2$. Initial point: $x_0 = [-1.2, 1]$. $f(x_0) = 24.2$. Solution: $x^* = [1, \ 1]$, $f(x^*) = 0$.

(18) *Freudenstein-Roth Function* (Andrei 2003, p. 41) [FRE-ROTH]

$$f(x) = \sum_{i=1}^{n/2} \left(-13 + x_{2i-1} + ((5 - x_{2i})x_{2i} - 2)x_{2i}\right)^2$$
$$+ \left(-29 + x_{2i-1} + ((x_{2i} + 1)x_{2i} - 14)x_{2i}\right)^2,$$

$n = 2$. Initial point: $x_0 = [0.5, -2., 0.5, -2., \ldots, 0.5, -2.]$. $f(x_0) = 400.5$. Solution: $x^* = [11.412836, -0.896802]$, $f(x^*) = 48.984253$.

(19) *White & Holst Function* (Andrei 2003, p. 42) [WHI-HOL]

$$f(x) = \sum_{i=1}^{n/2} c\left(x_{2i} - x_{2i-1}^3\right)^2 + \left(1 - x_{2i-1}\right)^2, \quad c = 1.$$

$n = 2$. Initial point: $x_0 = [-1.2, 1., \ldots, -1.2, 1.]$. $f(x_0) = 12.281984$. Solution: $x^* = [1, \ 1]$, $f(x^*) = 0$.

© The Author(s), under exclusive license to Springer Nature Switzerland AG 2021
N. Andrei, *A Derivative-free Two Level Random Search Method for Unconstrained Optimization*, SpringerBriefs in Optimization,
https://doi.org/10.1007/978-3-030-68517-1

(20) *Miele & Cantrell Function* [MI-CAN]

$$f(x) = (\exp(x_1) - x_2)^4 + 100(x_2 - x_3)^6 + \{\tan(x_3 - x_4)\}^4 + x_1^8.$$

$n = 4$. Initial point: $x_0 = [-1.2, 2, 2, 3]$. $f(x_0) = 13.008977$. Solution: $x^* = [0.0854, 1.0930, 1.1050, 1.1078]$, $f(x^*) = 0$.

(21) *Himmelblau Function* (Floudas, Pardalos et al. 1999, p. 326) [HIMM-1]

$$f(x) = (4x_1^3 + 4x_1x_2 + 2x_2^2 - 42x_1 - 14)^2 + (4x_2^3 + 2x_1^2 + 4x_1x_2 - 26x_2 - 22)^2.$$

$n = 2$. Initial point: $x_0 = [-1.2, 1]$. $f(x_0) = 2820.895744$. Solution: $x^* = [-0.127963, -1.953726]$, $f(x^*) = 0$.

(22) *Three-Hump Camel Back Function (1)* (Floudas, Pardalos et al. 1999, p. 111) [3-CAMEL]

$$f(x) = 4x_1^2 - 2.1x_1^4 + \frac{1}{3}x_1^6 + x_1x_2 - 4x_2^2 + 4x_2^4.$$

$n = 2$. Initial point: $x_0 = [1.2, 1]$. $f(x_0) = 3.600768$. Solution: $x^* = [0.0898414, -0.712658]$, $f(x^*) = -1.031628453$.

(23) *Six-Hump Camel Back Function (2)* (Floudas, Pardalos et al. 1999, p. 110) [6-CAMEL]

$$f(x) = 12x_1^2 - 6.3x_1^4 + x_1^6 - 6x_1x_2 + 6x_2^2.$$

$n = 2$. Initial point: $x_0 = [1, 1]$. $f(x_0) = 6.7$. Solution: $x^* = [0, 0]$, $f(x^*) = 0$.

(24) *Wood Function* (Andrei 2003, p. 42) [WOOD]

$$f(x) = \sum_{i=1}^{n/4} 100(x_{4i-3}^2 - x_{4i-2})^2 + (x_{4i-3} - 1)^2 + 90(x_{4i-1}^2 - x_{4i})^2$$
$$+ (1 - x_{4i-1})^2 + 10.1\{(x_{4i-2} - 1)^2 + (x_{4i} - 1)^2\} + 19.8(x_{4i-2} - 1)(x_{4i} - 1),$$

$n = 4$. Initial point: $x_0 = [-3, -1, -3, -1]$. $f(x_0) = 19192.0$. Solution: $x^* = [1, 1, 1, 1]$, $f(x^*) = 0$.

(25) *Sum of Quadratic Powers* [QUADR-2]

$$f(x) = (x_1 - 45)^2 + (x_2 - 90)^2 + 11.$$

$n = 2$. Initial point: $x_0 = [1.2, 1]$. $f(x_0) = 9850.44$. Solution: $x^* = [45, 90]$, $f(x^*) = 11$.

(26) *Shekel Function* (Floudas, Pardalos et al. 1999, p. 111) [SHEKEL]

$$f(x) = -\frac{1}{(x_1 - 4)^2 + (x_2 - 4)^2 + 0.1} - \frac{1}{(x_1 - 1)^2 + (x_2 - 1)^2 + 0.2}$$
$$- \frac{1}{(x_1 - 8)^2 + (x_2 - 8)^2 + 0.2}.$$

$n = 2$. Initial point: $x_0 = [1.2, \ 1]$. $f(x_0) = -4.236176$. Solution: $x^* = [4, \ 4]$, $f(x^*) = -10.0860014$.

(27) *DENSCHNA (CUTE)* (Gould et al. 2013) [DENSCHNA]

$$f(x) = \sum_{i=1}^{n/2} x_{2i-1}^4 + (x_{2i-1} + x_{2i})^2 + (-1 + \exp(x_{2i}))^2.$$

$n = 8$. Initial point: $x_0 = [1, \ 1, \ 1, \ 1, \ 1, \ 1, \ 1, \ 1]$. $f(x_0) = 31.809969$. Solution: $x^* = [0, \ 0, \ 0, \ 0, \ 0, \ 0, \ 0, \ 0]$, $f(x^*) = 0$.

(28) *DENSCHNB (CUTE)* (Gould et al. 2013) [DENSCHNB]

$$f(x) = \sum_{i=1}^{n/2} (x_{2i-1} - 2)^2 + (x_{2i-1} - 2)^2 (x_{2i})^2 + (x_{2i} + 1)^2.$$

$n = 2$. Initial point: $x_0 = [10, \ 10]$. $f(x_0) = 6585.0$. Solution: $x^* = [2, \ -1]$, $f(x^*) = 0$.

(29) *DENSCHNC (CUTE)* (Gould et al. 2013) [DENSCHNC]

$$f(x) = \sum_{i=1}^{n/2} \left(-2 + x_{2i-1}^2 + x_{2i}^2\right)^2 + \left(-2 + \exp(x_{2i-1} - 1) + x_{2i}^3\right)^2.$$

$n = 8$. Initial point: $x_0 = [3, \ 3, \ 3, \ 3, \ 3, \ 3, \ 3, \ 3]$. $f(x_0) = 5220.203819$. Solution: $x^* = [1, 1, 1, 1, 1, 1, 1, 1]$, $f(x^*) = 0$.

(30) *Griewank Function* (Jamil and Yang 2013) [GRIEWANK]

$$f(x) = 1 + \frac{x_1^2 + x_2^2}{4000} - \cos(x_1) \cos\left(\frac{x_2}{\sqrt{2}}\right).$$

$n = 2$. Initial point: $x_0 = [-1.2, \ 1]$. $f(x_0) = 0.725129$. Solution: $x^* = [0, \ 0]$, $f(x^*) = 0$.

(31) *Brent Function* (Brent 1972) [BRENT]

$$f(x) = (x_1 + 10)^2 + (x_2 + 10)^2 - \exp\left(-x_1^2 - x_2^2\right).$$

$n = 2$. Initial point: $x_0 = [0, \ 0]$. $f(x_0) = 199.0$. Solution: $x^* = [-10, \ -10]$, $f(x^*) = 0$.

(32) *Booth Function* (Jamil and Yang 2013) [BOOTH]

$$f(x) = (x_1 + 2x_2 - 7)^2 + (2x_1 + x_2 - 5)^2.$$

$n = 2$. Initial point: $x_0 = [1.2, \ 1]$. $f(x_0) = 17.0$. Solution: $x^* = [1, \ 3]$, $f(x^*) = 0$.

(33) *Matyas Function* (Jamil and Yang 2013) [MATYAS]

$$f(x) = 0.28(x_1^2 + x_2^2) - 0.48x_1x_2,$$

$n = 2$. Initial point: $x_0 = [1.2, \ 1]$. $f(x_0) = 0.1072$. Solution: $x^* = [0, \ 0]$, $f(x^*) = 0$.

(34) *Colville Function* (Andrei 2003, p. 44) [COLVILLE]

$$f(x) = 100\left(x_3^2 + \left(x_2 - x_1^2\right)^2\right) + (1 - x_1)^2 + (x_3 - x_1 + x_2)^2$$
$$+ 10.1\left((x_2 - 1)^2 + (x_4 - 1)^2\right) + 19.8(x_2 - 1)(x_4 - 1).$$

$n = 3$. Initial point: $x_0 = [10, \ 2, \ 3]$. $f(x_0) = 961528.04$. Solution: $x^* = [1, \ 1, \ 0]$, $f(x^*) = 0$.

(35) *Easom Function* (Jamil and Yang 2013) [EASOM]

$$f(x) = -\cos(x_1)\cos(x_2)\exp\left(-(x_1 - \pi)^2 - (x_2 - \pi)^2\right).$$

$n = 2$. Initial point: $x_0 = [1, \ 1]$. $f(x_0) = -0.303089E - 04$. Solution: $x^* = [\pi, \ \pi]$, $f(x^*) = -1$. ($\pi = 3.14159094822$)

(36) *Beale Function* (Andrei 2003, p. 42) [BEALE]

$$f(x) = \sum_{i=1}^{n/2}(1.5 - x_{2i-1}(1 - x_{2i}))^2 + \left(2.25 - x_{2i-1}\left(1 - x_{2i}^2\right)\right)^2$$
$$+ \left(2.625 - x_{2i-1}\left(1 - x_{2i}^3\right)\right)^2,$$

$n = 8$. Initial point: $x_0 = [1., 0.8, \ldots, 1., 0.8]$. $f(x_0) = 39.315476$. Solution: $x^* = [3, \ 0.5, \ 3, \ 0.5, \ 3, \ 0.5, \ 3, \ 0.5]$, $f(x^*) = 0$.

(37) *Powell Function* (Andrei 2003, p. 41) [POWELL]

$$f(x) = (x_1 + 10x_2)^2 + 5(x_3 - x_4)^2 + (x_2 - 2x_3)^4 + 10(x_1 - x_4)^4.$$

$n = 4$. Initial point: $x_0 = [3, \ -1, \ 0, \ 1]$. $f(x_0) = 215.0$. Solution: $x^* = [0, \ 0, \ 0, \ 0]$, $f(x^*) = 0$.

(38) *McCormick Function* (Jamil and Yang 2013) [McCORM]

$$f(x) = \sin(x_1 + x_2) + (x_1 - x_2)^2 - 1.5x_1 + 2.5x_2 + 1.$$

$n = 2$. Initial point: $x_0 = [-2, \ 0]$. $f(x_0) = 7.090702$. Solution: $x^* = [-0.547198, \ -1.5472005]$, $f(x^*) = -1.9132229$.

(39) *Himmelblau Function (-11, -7)* (Andrei 2003, p. 41) [HIMM-2]

$$f(x) = \left(x_1^2 + x_2 - 11\right)^2 + \left(x_1 + x_2^2 - 7\right)^2.$$

$n = 2$. Initial point: $x_1 = [1, \ 1]$. $f(x_0) = 106.0$. Solution: $x^* = [3, \ 2]$, $f(x^*) = 0$.

(40) *Leon Function* (Jamil and Yang 2013) [LEON]

$$f(x) = 100\left(x_2 - x_1^3\right)^2 + (x_1 - 1)^2.$$

$n = 2$. Initial point: $x_0 = [-1.2, \ 1]$. $f(x_0) = 749.0384$. Solution: $x^* = [1, \ 1]$, $f(x^*) = 0$.

(41) *Price Function No.4* (Jamil and Yang 2013) [PRICE4]

$$f(x) = \left(2x_1^3 x_2 - x_2^3\right)^2 + \left(6x_1 - x_2^2 - x_2\right)^2.$$

$n = 2$. Initial point: $x_0 = [1.2, \ 1]$. $f(x_0) = 33.071936$. Solution: $x^* = [0.0018, \ 0.0111]$, $f(x^*) = 0$.

(42) *Zettl Function* (Jamil and Yang 2013) [ZETTL]

$$f(x) = \left(x_1^2 + x_2^2 - 2x_1\right)^2 + 0.25x_1.$$

$n = 2$. Initial point: $x_0 = [1, \ 1]$. $f(x_0) = 25.0$. Solution: $x^* = [-0.029896, \ 0]$, $f(x^*) = -0.00379123$.

(43) *Sphere Function* (Jamil and Yang 2013) [SPHERE]

$$f(x) = \sum_{i=1}^{n} (x_i - 1)^2.$$

$n = 8$. Initial point: $x_0 = [10, \ldots, 10]$. $f(x_0) = 648.0$. Solution: $x^* = [1, \ldots, 1]$. $f(x^*) = 0$.

(44) *Ellipsoid Function* [ELLIPSOID]

$$f(x) = \sum_{i=1}^{n} (x_i - i)^2.$$

$n = 8$. Initial point: $x_0 = [0, \ldots, 0]$. $f(x_0) = 204.0$. Solution: $x^* = [1, 2, 3, 4, 5, 6, 7, 8]$, $f(x^*) = 0$.

(45) *Himmelblau Function (Problem 29, p. 428)* (Himmelblau 1972, p. 428) [HIMM-3]

$$f(x) = \left(x_1^2 + 12x_2 - 1\right)^2 + \left(49x_1^2 + 49x_2^2 + 84x_1 + 2324x_2 - 681\right)^2.$$

$n = 2$. Initial point: $x_0 = [1, \ 1]$. $f(x_0) = 3330769.0$. Solution: $x^* = [0.285969, \ 0.279318]$, $f(x^*) = 5.922563$.

(46) *Himmelblau Function (Problem 30, p. 428)* (Himmelblau 1972, p. 428) [HIMM-4]

$$f(x) = 100\left(x_3 - \left(\frac{x_1 + x_2}{2}\right)^2\right)^2 + (1 - x_1)^2 + (1 - x_2)^2.$$

$n = 3$. Initial point: $x_0 = [-1.2, \ 2., 0.]$. $f(x_0) = 84.0$. Solution: $x^* = [1, \ 1, \ 1]$, $f(x^*) = 0$.

(47) *Himmelblau Function (Problem 33, p. 430)* (Himmelblau 1972, p. 430) [HIMM-5]

$$f(x) = \left(2x_1^2 + 3x_2^2\right)\exp\left(-x_1 - x_2\right).$$

$n = 2$. Initial point: $x_0 = [0.5, \ 0.5]$. $f(x_0) = 0.459849$. Solution: $x^* = [0., \ 0.], f(x^*) = 0$.

(48) *Zirilli Function* (Jamil and Yang 2013) [ZIRILLI]

$$f(x) = 0.25x_1^4 - 0.5x_1^2 + 0.1x_1 + 0.5x_2^2.$$

$n = 2$. Initial point: $x_0 = [0, \ 4]$. $f(x_0) = 8.0$. Solution: $x^* = [-1.046681, 0]$, $f(x^*) = -0.352386$.

(49) *Styblinski Function* (Jamil and Yang 2013) [STYBLIN]

$$f(x) = \left(x_1^4 - 16x_1^2 + 5x_1\right)/2 + \left(x_2^4 - 16x_2^2 + 5x_2\right)/2.$$

$n = 2$. Initial point: $x_0 = [-1, \ -1]$. $f(x_0) = -20.0$. Solution: $x^* = [-2.903537, \ -2.903537]$, $f(x^*) = -78.332331$.

(50) *Trid Function* (Jamil and Yang 2013) [TRID]

$$f(x) = (x_1 - 1)^2 + (x_2 - 1)^2 - x_1 x_2.$$

$n = 2$. Initial point: $x_0 = [-1, \ -1]$. $f(x_0) = 7.0$. Solution: $x^* = [2, \ 2], f(x^*) = -2$.

(51) *A Scaled Quadratic Function:* [SCALQ]

$$f(x) = 10000(x_1 - 10)^2 + 0.00001(x_2 - 5)^2.$$

$n = 2$. Initial point: $x_0 = [1, \ 1]$. $f(x_0) = 810000.0$. Solution: $x^* = [10, \ 5.025615]$, $f(x^*) = 0$.

(52) *A Three-Variable Problem* (Schittkowski 1987, Problem 241, p. 65) [SCHIT-1]

$$f(x) = \left(x_1^2 + x_2^2 + x_3^2 - 1\right)^2$$
$$+\left(x_1^2 + x_2^2 + (x_3 - 2)^2 - 1\right)^2$$
$$+(x_1 + x_2 + x_3 - 1)^2$$
$$+(x_1 + x_2 - x_3 + 1)^2$$
$$+\left(x_1^3 + 3x_2^2 + (5x_3 - x_1 + 1)^2 - 36\right)^2.$$

$n = 3$. Initial point: $x_0 = [1, \ 2, \ 0]$. $f(x_0) = 629.0$. Solution: $x^*[0, \ 0, \ 1]$, $f(x^*) = 0$.

(53) *A Six-Variable Problem* (Schittkowski 1987, Problem 271, p. 95) [SCHIT-2]

$$f(x) = \sum_{i=1}^{6} (16 - i)(x_i - 1)^2.$$

$n = 6$. Initial point: $x_0 = [0, 0, 0, 0, 0, 0]$. $f(x_0) = 75.0$. Solution:
$x^* = [1, 1, 1, 1, 1, 1]$, $f(x^*) = 0$.

(54) *A Two-Variable Problem* (Schittkowski 1987, Problem 308, p. 131) [SCHIT-3]

$$f(x) = \left(x_1^2 + x_2^2 + x_1 x_2\right)^2 + \sin^2 x_1 + \cos^2 x_2.$$

$n = 2$. Initial point: $x_0 = [3, \ 0.1]$. $f(x_0) = 87.686048$.
Solution: $x^* = [0.155438, \ -0.694560]$. $f(x^*) = 0.773199$.

(55) *Brown's Almost Linear System* (Floudas, Pardalos et al. 1999, p. 329) [BROWN-A]

$$f(x) = (2x_1 + x_2 + x_3 + x_4 + x_5 - 6)^2$$
$$+(x_1 + 2x_2 + x_3 + x_4 + x_5 - 6)^2$$
$$+(x_1 + x_2 + 2x_3 + x_4 + x_5 - 6)^2$$
$$+(x_1 + x_2 + x_3 + 2x_4 + x_5 - 6)^2$$
$$+(x_1 x_2 x_3 x_4 x_5 - 1)^2.$$

$n = 5$. Initial point: $x_0 = [0.5, \ldots, 0.5]$. $f(x_0) = 36.938476$. Solution: $x^* = [1, 1, 1, 1, 1]$, $f(x^*) = 0$.

(56) *Kelley Function* (Andrei 2003, p. 59) [KELLEY]

$$f(x) = (x_1 - x_2 x_3 x_4)^2 + (x_2 - x_3 x_4)^2 + (x_3 - x_4)^2 + x_4^2.$$

$n = 4$. Initial point: $x_0 = [2, \ 2, \ 2, \ 2]$. $f(x_0) = 44.0$. Solution: $x^* = [0, \ 0, \ 0, \ 0]$,
$f(x^*) = 0$.

(57) *A Nonlinear System* (Andrei 2003, p. 59) [NONSYS]

$$f(x) = \left(8 - x_1^2 - x_2^2 - x_3^2 - x_4^2 - x_1 + x_2 - x_3 + x_4\right)^2$$
$$+ \left(10 - x_1^2 - 2x_2^2 - x_3^2 - 2x_4^2 + x_1 + x_4\right)^2$$
$$+ \left(5 - 2x_1^2 - x_2^2 - x_3^2 - 2x_1 + x_2 + x_4\right)^2.$$

$n = 4$. Initial point: $x_0 = [0, \ 0, \ 0, \ 0]$. $f(x_0) = 189.0$.
Solution: $x^* = [-0.329153, \ -1.740923, \ 1.363694, \ 1.1897362]$, $f(x^*) = 0$.

(58) *Zangwill Function* (Andrei 2003, p. 42) [ZANGWILL]

$$f(x) = \frac{1}{15}\left(16x_1^2 + 16x_2^2 - 8x_1x_2 - 56x_1 - 256x_2 + 991\right).$$

$n = 2$. Initial point: $x_0 = [3, \ 8]$. $f(x_0) = -16.60$. Solution: $x^* = [4, \ 9]$, $f(x^*) = -18.2$.

(59) *Circular Function* (Andrei 2003, p. 43) [CIRCULAR]

$$f(x) = x_1 + x_2 + x_3 + x_1x_2 + x_2x_3 + x_3x_1 + x_1x_2x_3 + x_1^2 + x_2^2 + x_3^2.$$

$n = 3$. Initial point: $x_0 = [1, \ 1, \ 1]$. $f(x_0) = 10.0$.
Solution: $x^* = [-0.2679439, \ -0.2679417, \ -0.26795721]$, $f(x^*) = -0.392304845$.

(60) *Polexp Function* (Andrei 2003, p. 43) [POLEXP]

$$f(x) = \left(1 + x_1 + x_2 + x_1x_2 + x_1^2 + x_2^2 + x_1^3 + x_2^3\right)\exp(-x_1)\exp(-x_2).$$

$n = 2$. Initial point: $x_0 = [1, \ 1]$. $f(x_0) = 1.082682$.
Solution: $x^* = [11.171158, \ 11.531956]$, $f(x^*) = 0$.

(61) *Dulce Function* (Andrei 2003, p. 43) [DULCE]

$$f(x) = \frac{(x_1 + 1)x_2}{(x_1 + 1)^2 + x_2^2}.$$

$n = 2$. Initial point: $x_0 = [1, \ 1]$. $f(x_0) = 4.0$.
Solution: $x^* = [-13.377186, \ 12.377186]$, $f(x^*) = -0.5$.

(62) *Cragg & Levy Function* (Andrei 2003, p. 49) [CRAGLEVY]

$$f(x) = \left(\exp(x_1) - x_2\right)^4 + 100(x_2 - x_3)^6 + (tg(x_3 - x_4))^4 + x_1^8 + (x_4 - 1)^2.$$

$n = 4$. Initial point: $x_0 = [1, \ 2, \ 2, \ 2]$. $f(x_0) = 2.266182$. Solution: $x^* = [0, \ 1, \ 1, \ 1]$, $f(x^*) = 0$.

(63) *Broyden Tridiagonal Function, n = 5.* (Andrei 2003, p. 50) [BROYDEN1]

$$f(x) = ((3 - 2x_1)x_1 - 2x_2 + 1)^2$$
$$+ \sum_{i=2}^{n-1} ((3 - 2x_i)x_i - x_{i-1} - 2x_{i+1} + 1)^2$$
$$+ ((3 - 2x_n)x_n - x_{n-1} + 1)^2.$$

$n = 5$. Initial point: $x_0 = [-1, -1, -1, -1, -1]$. $f(x_0) = 16.0$. Solution: $x^* = [1.821044, -0.845943, -0.544650, -0.571356, -0.411105], f(x^*) = 0$.

(64) *Broyden Tridiagonal Function, n = 8* (Andrei 2003, p. 50) [BROYDEN2]
$n = 8$. Initial point: $x_0 = [-1, \ldots, -1]$. $f(x_0) = 19.0$.
Solution:

$x(1) = 0.1648879791010E + 00$ $x(5) = 0.3673718041848E + 00$

$x(2) = 0.7635517844971E + 00$ $x(6) = 0.7738330712177E + 00$

$x(3) = 0.1081441217480E + 01$ $x(7) = 0.1188326976550E + 01$

$x(4) = 0.4785779873528E + 00$ $x(8) = 0.2598464652005E + 00$

$f(x^*) = 0.939397$.

(65) *Broyden Tridiagonal Function, n = 10* (Andrei 2003, p. 50) [BROYDEN3]
$n = 10$. Initial point: $x_0 = [-1, \ldots, -1]$. $f(x_0) = 21.0$.
Solution:

$x(1) = -0.4530890768552E + 00$ $x(6) = 0.4215137979055E + 00$

$x(2) = -0.3827293425644E + 00$ $x(7) = 0.3105819051529E + 00$

$x(3) = 0.1660023233420E - 01$ $x(8) = 0.7592162159579E + 00$

$x(4) = 0.7597511735344E + 00$ $x(9) = 0.1210457174025E + 01$

$x(5) = 0.1161414813843E + 01$ $x(10) = 0.2587657464974E + 00$

$f(x^*) = 0.878973$.

(66) *Full Rank, n = 5.* (Andrei 2003, p. 51) [FULRANK1]

$$f(x) = (x_1 - 3)^2 + \sum_{i=2}^{n} \left(x_1 - 3 - 2(x_1 + x_2 + \cdots + x_i)^2 \right)^2.$$

$n = 5$. Initial point: $x_0 = [-1, \ldots, -1]$. $f(x_0) = 4856.0$. Solution: $x^* = [3, -3, 0.000057, -0.0012, 0.0014], f(x^*) = 0$.

(67) *Full Rank, n = 8.* (Andrei 2003, p. 51) [FULRANK2]
$n = 8$. Initial point: $x_0 = [-1, \ldots, -1]$. $f(x_0) = 38460.0$.
Solution:

$x(1) = 0.3000929327372E + 01$ $x(5) = 0.5171399472856E - 01$

$x(2) = -0.3016792870409E + 01$ $x(6) = 0.2736696756323E - 02$

$x(3) = 0.3229127177846E - 01$ $x(7) = -0.4509531133148E - 01$

$x(4) = -0.4332983920218E - 01$ $x(8) = -0.9535619626329E - 02$

$f(x^*) = 0$.

(68) *Full Rank, n = 10.* (Andrei 2003, p. 51) [FULRANK3]
$n = 10$. Initial point: $x_0 = [-1, \ldots, -1]$. $f(x_0) = 107632.0$.
Solution:

$x(1) = 0.3000884879922E + 01$	$x(6) = 0.2568205976576E - 01$
$x(2) = -0.3022100757867E + 01$	$x(7) = -0.4698446625383E - 01$
$x(3) = -0.3552470964438E - 02$	$x(8) = 0.2782783115725E - 02$
$x(4) = 0.4859580696293E - 01$	$x(9) = 0.4755699668969E - 01$
$x(5) = -0.2688681079712E - 01$	$x(10) = -0.5113547437530E - 01$

$f(x^*) = 0$.

(69) *Trigonometric, n = 5.* (Andrei 2003, p. 51) [TRIG-1]

$$f(x) = \sum_{i=1}^{n} \left(\left(n - \sum_{j=1}^{n} \cos(x_j) \right) + i(1 - \cos(x_i) - \sin(x_i)) \right)^2.$$

$n = 5$. Initial point: $x_0 = [0.2, \ldots, 0.2]$. $f(x_0) = 0.0116573$. Solution: $x^* = [0, 0, 0, 0, 0]$, $f(x^*) = 0$.

(70) *Trigonometric, n = 9.* (Andrei 2003, p. 51) [TRIG-2]
$n = 9$. Initial point: $x_0 = [0.2, \ldots, 0.2]$. $f(x_0) = 0.0820158$. Solution: $x^* = [0, \ldots, 0]$, $f(x^*) = 0$.

(71) *Trigonometric, n = 20.* (Andrei 2003, p. 51) [TRIG-3]
$n = 20$. Initial point: $x_0 = [0.2, \ldots, 0.2]$. $f(x_0) = 3.614762$. Solution: $x^* = [0, \ldots, 0]$, $f(x^*) = 0$.

(72) *Brown Function* (Andrei 2003, p. 59) [BROWN]

$$f(x) = \sum_{i=1}^{n-1} \left(x_i^2 \right)^{\left(x_{i+1}^2 + 1 \right)} + \left(x_{i+1}^2 \right)^{\left(x_i^2 + 1 \right)}.$$

$n = 10$. Initial point: $x_0 = [0.1, \ldots, 0.1]$. $f(x_0) = 0.171898$. Solution: $x^* = [0, \ldots, 0]$, $f(x^*) = 0$.

(73) *Brown and Dennis Function* (Andrei 2003, p. 46) [BRODEN]

$$f(x) = \sum_{i=1}^{20} \left((x_1 + t_i x_2 - \exp(t_i))^2 + (x_3 + x_4 \sin(t_i) - \cos(t_i))^2 \right)^2,$$

$$t_i = i/5, \quad i = 1, \ldots, 20.$$

$n = 4$. Initial point: $x_0 = [25, \ 5, \ -5, \ -1]$. $f(x_0) = 7926693.336$.
Solution:

$x(1) = -0.1159447239572E + 02$	$x(3) = -0.4034427663215E + 00$
$x(2) = 0.1320364113202E + 02$	$x(4) = 0.2367521691007E + 00$

$f(x^*) = 85822.20162847$.

(74) *Hosaki Function* (Moré et al. 1981) [HOSAKI]

$$f(x) = \left(1 - 8x_1 + 7x_1^2 - \frac{7}{3}x_1^3 + \frac{1}{4}x_1^4\right)x_2^2 \exp(-x_2).$$

$n = 2$. Initial point: $x_0 = [1, \ 1]$. $f(x_0) = -0.766415$. Solution: $x^* = [4, \ 2]$, $f(x^*) = -2.3458$.

(75) *Cosmin Function:* [COSMIN]

$$f(x) = (|x_1| - 5)^2 + (|x_2| - 8)^2.$$

$n = 2$. Initial point: $x_0 = [1, \ 1]$. $f(x_0) = 65.0$. Solution: $x^* = [-5, \ -8]$, $f(x^*) = 0$.

(76) *BDQRTIC (CUTE)* (Gould et al. 2013; Andrei 2020, p. 461) [BDQRTIC]

$$f(x) = \sum_{i=1}^{n-4} (-4x_i + 3)^2 + \left(x_i^2 + 2x_{i+1}^2 + 3x_{i+2}^2 + 4x_{i+3}^2 + 5x_n^2\right)^2,$$

$n = 16$. Initial point: $x_0 = [1., 1., \ldots, 1.]$. $f(x_0) = 2712.0$.
Solution:

$x(1) = 0.6249919832552E + 00 \quad x(9) = 0.3211986118300E + 00$
$x(2) = 0.4852161180549E + 00 \quad x(10) = 0.3230094832868E + 00$
$x(3) = 0.3712664183308E + 00 \quad x(11) = 0.3364175022838E + 00$
$x(4) = 0.2858607826828E + 00 \quad x(12) = 0.3743689759758E + 00$
$x(5) = 0.3182696323183E + 00 \quad x(13) = -0.1906881996559E - 03$
$x(6) = 0.3266637280239E + 00 \quad x(14) = -0.1005917908602E - 03$
$x(7) = 0.3259000009599E + 00 \quad x(15) = -0.2420039722632E - 02$
$x(8) = 0.3239842465587E + 00 \quad x(16) = -0.6855008112671E - 04$

$f(x^*) = 42.297914$.

(77) *DIXON3DQ (CUTE)* (Gould et al. 2013; Andrei 2020, p. 463) [DIXON3DQ]

$$f(x) = (x_1 - 2)^2 + \sum_{i=1}^{n-1} (x_i - x_{i+1})^2 + (x_n - 1)^2,$$

$n = 10$. Initial point: $x_0 = [-1, -1, \ldots, -1]$. $f(x_0) = 13.0$.
Solution:

$x(1) = 0.1906610914742E + 01 \quad x(6) = 0.1444987776382E + 01$
$x(2) = 0.1813194277947E + 01 \quad x(7) = 0.1355225454058E + 01$
$x(3) = 0.1720110818585E + 01 \quad x(8) = 0.1266071728416E + 01$
$x(4) = 0.1627765880424E + 01 \quad x(9) = 0.1177492321262E + 01$
$x(5) = 0.1535948879928E + 01 \quad x(10) = 0.1088654808627E + 01$

$f(x^*) = 0.0909483$.

(78) *ENGVAL1 (CUTE)* (Gould et al. 2013; Andrei 2020, p. 462) [ENGVAL1]

$$f(x) = \sum_{i=1}^{n-1} \left(x_i^2 + x_{i+1}^2\right)^2 + \sum_{i=1}^{n-1} (-4x_i + 3),$$

$n = 4$. Initial point: $x_0 = [2.,2.,\ldots,2.]$. $f(x_0) = 177.0$.
  Solution:

$x(1) = 0.9093567644139E + 00$ $x(3) = 0.7365353093242E + 00$

$x(2) = 0.5222240376987E + 00$ $x(4) = -0.4540210073353E - 04$

$f(x^*) = 2.4956043$.

(79) *Extended Penalty Function* (Andrei 2020, p. 459) [EX-PEN]

$$f(x) = \left(\sum_{i=1}^{n} x_i^2 - 0.25\right)^2 + \sum_{i=1}^{n-1} (x_i - 1)^2,$$

$n = 5$. Initial point: $x_0 = [1, 2, \ldots, n]$. $f(x_0) = 3011.5625$.
  Solution:

$x(1) = 0.4584615600207E + 00$ $x(4) = 0.4584551430102E + 00$

$x(2) = 0.4584263438130E + 00$ $x(5) = 0.3593922907023E - 05$

$x(3) = 0.4584087207781E + 00$

$f(x^*) = 1.52203872$.

(80) *Broyden Pentadiagonal Function.* (Andrei 2020, p. 458) [BROYDENP]

$$f(x) = \left(3x_1 - 2x_1^2\right)^2 + \sum_{i=2}^{n-1} \left(3x_i - 2x_i^2 - x_{i-1} - 2x_{i+1} + 1\right)^2$$
$$+ \left(3x_n + 2x_n^2 - x_{n-1} + 1\right)^2,$$

$n = 5$. Initial point: $x_0 = [-1., -1., \ldots, -1.]$. $f(x_0) = 37.0$.
  Solution:

$x(1) = -0.2095006858067E - 01$ $x(4) = 0.1284819134763E + 01$

$x(2) = 0.8037324362144E - 01$ $x(5) = 0.2469520183489E + 00$

$x(3) = 0.7228876152828E + 00$

$f(x^*) = 0.544499$.

(81) *Teo Function:* [TEO]

$$f(x) = 1 + \sin^2 x_1 + \sin^2 x_2 - 0.1 \exp\left(-x_1^2 - x_2^2\right).$$

$n = 2$. Initial point: $x_0 = [1, \ 1]$. $f(x_0) = 2.402613$. Solution: $x^* = [0, 0]$, $f(x^*) = 0.9$.

(82) *Coca Function:* [COCA]

$$f(x) = 100\left(x_2 - x_1^2\right)^2 + \left(6.4(x_2 - 0.5)^2 - x_1 - 0.6\right)^2.$$

$n = 2$. Initial point: $x_0 = [2.5, \ 2.5]$. $f(x_0) = 1912.5$.
Solution: $x^* = [-0.663678, \ 0.441117]$, $f(x^*) = 0.00741539$.

(83) *NEC Function* (Andrei 2003, p. 46) [NEC]

$$f(x) = 2x_1^2 + x_2^2 + 2x_1 x_2 + x_1 - x_2.$$

$n = 2$. Initial point: $x_0 = [1, \ 1]$. $f(x_0) = 5.0$. Solution: $x^* = [-1, \ 1.5]$, $f(x^*) = 1.25$.

(84) *QuadraticPowerExp* (Andrei 2003, p. 52) [QPEXP]

$$f(x) = (x_1 x_2 x_3 x_4)^2 \exp\left(-x_1 - x_2 - x_3 - x_4\right).$$

$n = 4$. Initial point: $x_0 = [3, \ 4, \ 0.5, \ 1]$. $f(x_0) = 0.00732486$.
Solution:

$$x(1) = 0.8439804730716E + 01 \quad x(3) = -0.2235174179069E - 07$$
$$x(2) = 0.6106801441938E + 01 \quad x(4) = 0.1797781093531E + 01$$

$$f(x^*) = 0.$$

(85) *NONDQAR (CUTE)* (Gould et al. 2013; Andrei 2020, p. 461) [NONDQAR]

$$f(x) = (x_1 - x_2)^2 + \sum_{i=1}^{n-2} (x_i + x_{i+1} + x_n)^4 + (x_{n-1} + x_n)^2,$$

$n = 8$. Initial point: $x_0 = [1, \ldots, 1]$. $f(x_0) = 490.0$.
Solution:

$$x(1) = -0.1068348857385E - 03 \quad x(5) = -0.1466339867001E - 02$$
$$x(2) = -0.1030386928212E - 03 \quad x(6) = -0.9392189280798E - 03$$
$$x(3) = 0.4689700201423E - 03 \quad x(7) = -0.8870365625862E - 03$$
$$x(4) = -0.8716260923858E - 03 \quad x(8) = 0.8819461036838E - 03$$

$$f(x^*) = 0.$$

(86) *ARWHEAD (CUTE)* (Gould et al. 2013; Andrei 2020, p. 461) [ARWHEAD]

$$f(x) = \sum_{i=1}^{n-1} (-4x_i + 3) + \sum_{i=1}^{n-1} \left(x_i^2 + x_n^2\right)^2,$$

$n = 40$. Initial point: $x_0 = [1., 1., \ldots, 1.]$. $f(x_0) = 117.0$. Solution: $x^* = [1, 1, \ldots, 1, 0]$, $f(x^*) = 0$.

(87) *CUBE (CUTE)* (Gould et al. 2013; Andrei 2020, p. 466) [CUBE]

$$f(x) = (x_1 - 1)^2 + \sum_{i=2}^{n} 100\left(x_i - x_{i-1}^3\right)^2,$$

$n = 4$. Initial point: $x_0 = [-1.2, \ 1, \ldots, -1.2, \ 1]$. $f(x_0) = 1977.237$. Solution: $x^* = [1, \ 1, \ 1, \ 1]$, $f(x^*) = 0$.

(88) *NONSCOMP (CUTE)* (Gould et al. 2013; Andrei 2020, p. 466) [NONSCOMP]

$$f(x) = (x_1 - 1)^2 + \sum_{i=2}^{n} 4\left(x_i - x_{i-1}^2\right)^2,$$

$n = 5$. Initial point: $x_0 = [3., 3., \ldots, 3.]$. $f(x_0) = 580.0$. Solution: $x^* = [1, \ 1, \ 1, \ 1, \ 1]$, $f(x^*) = 0$.

(89) *DENSCHNF (CUTE)* (Gould et al. 2013; Andrei 2020, p. 463) [DENSCHNF]

$$f(x) = \sum_{i=1}^{n/2} \left(2(x_{2i-1} + x_{2i})^2 + (x_{2i-1} - x_{2i})^2 - 8\right)^2 + \left(5x_{2i-1}^2 + (x_{2i} - 3)^2 - 9\right)^2,$$

$n = 10$. Initial point: $x_0 = [2, \ 0, \ \ldots, \ 2, \ 0]$. $f(x_0) = 2080.0$. Solution: $x^* = [1, \ \ldots, \ 1]$, $f(x^*) = 0$.

(90) *BIGGSB1 (CUTE)* (Gould et al. 2013; Andrei 2020, p. 463) [BIGGSB1]

$$f(x) = (x_1 - 1)^2 + (1 - x_n)^2 + \sum_{i=2}^{n} (x_i - x_{i-1})^2,$$

$n = 16$. Initial point: $x_0 = [0, 0, \ldots, 0]$. $f(x_0) = 2.0$. Solution: $x^* = [1, \ \ldots, 1]$, $f(x^*) = 0$.

(91) *Borsec6:* [BORSEC6]

$$f(x) = \sum_{i=1}^{n} x_i^2 + \left(\frac{1}{2}\sum_{i=1}^{n} ix_i\right)^2 + \left(\frac{1}{4}\sum_{i=1}^{n} ix_i\right)^4 + \left(\frac{1}{6}\sum_{i=1}^{n} ix_i\right)^6.$$

$n = 30$. Initial point: $x_0 = [1, 1, \ldots, 1]$. $f(x_0) = 0.216858E + 12$. Solution: $x^* = [0, \ldots, 0]$, $f(x^*) = 0$.

(92) *Three Terms All Quadratics:* [3T-QUAD]

$$f(x) = \left(x_1^2 + x_2 - 10\right)^2 + \left(x_1 + x_2^2 - 7\right)^2 + \left(x_1^2 + x_2^3 - 70\right)^2.$$

$n = 2$. Initial point: $x_0 = [100, 100]$. $f(x_0) = 0.102016E + 13$. Solution: $x^* = [-2.8388596, 3.9352353]$, $f(x^*) = 36.867571$.

(93) *Mishra9* (Jamil and Yang 2013) [MISHRA9]

$$f(x) = \left(ab^2c + abc^2 + b^2 + d^2\right)^2,$$

$a = 2x_1^3 + 5x_1x_2 + 4x_3 - 2x_1^2x_3 - 18$, $\quad b = x_1 + x_2^3 + x_1x_3^2 - 22$,
$c = 8x_1^2 + 2x_2x_3 + 2x_2^2 + 3x_2^3 - 52$, $\quad d = x_1 + x_2 - x_3$

$n = 3$. Initial point: $x_0 = [10, 10, 10]$. $f(x_0) = 0.697625E + 27$.

Solution:
$$x(1) = 0.1529249316619E + 01$$
$$x(2) = 0.2197571947655E + 01$$
$$x(3) = -0.2459642162581E + 01$$

$f(x^*) = 0$.

(94) *Wayburn1* (Jamil and Yang 2013) [WAYBURN1]

$$f(x) = \left(x_1^6 + x_2^4 - 17\right)^2 + (2x_1 + x_2 - 4)^2.$$

$n = 2$. Initial point: $x_0 = [3, 3]$. $f(x_0) = 628874.0$. Solution: $x^* = [1.5968, 0.80644]$, $f(x^*) = 0$.

(95) *Wayburn2* (Jamil and Yang 2013) [WAYBURN2]

$$f(x) = \left(1.613 - 4(x_1 - 0.3125)^2 - 4(x_2 - 1.625)^2\right)^2 + (x_2 - 1)^2.$$

$n = 2$. Initial point: $x_0 = [2, 2]$. $f(x_0) = 107.918185$. Solution: $x^* = [0.2, 1]$, $f(x^*) = 0$.

(96) *Dixon & Price* (Jamil and Yang 2013) [DIX&PRI]

$$f(x) = (x_1 - 1)^2 + \sum_{i=2}^{n} i\left(2x_i^2 - x_{i-1}\right)^2.$$

$n = 10$. Initial point: $x_0 = [5, \ldots, 5]$. $f(x_0) = 109366.0$.

Solution:

$x(1) = 0.3333657322962E + 00$ $x(6) = -0.3931882592423E - 04$

$x(2) = -0.1159341387520E - 03$ $x(7) = 0.1978391149084E - 04$

$x(3) = -0.2738576797470E - 04$ $x(8) = 0.6564913302944E - 03$

$x(4) = 0.4707988846269E - 04$ $x(9) = 0.1789171446121E - 01$

$x(5) = -0.7333936236569E - 04$ $x(10) = 0.9450656511887E - 01$

$f(x^*) = 0.6666666.$

(97) *Qing* (Jamil and Yang 2013) [QING]

$$f(x) = \sum_{i=1}^{n} \left(x_i^2 - i \right)^2.$$

$n = 15$. Initial point: $x_0 = [2, \ldots, 2]$. $f(x_0) = 520.0$.

Solution:

$x(1) = 0.1000170830285E + 01$ $x(9) = 0.3000019489636E + 01$

$x(2) = 0.1414213034338E + 01$ $x(10) = 0.3162305277618E + 01$

$x(3) = -0.1732171266845E + 01$ $x(11) = 0.3316579442796E + 01$

$x(4) = 0.2000031446169E + 01$ $x(12) = 0.3464140551649E + 01$

$x(5) = 0.2236031832636E + 01$ $x(13) = 0.3605544042035E + 01$

$x(6) = 0.2449507746330E + 01$ $x(14) = -0.3741615293331E + 01$

$x(7) = 0.2645742732792E + 01$ $x(15) = -0.3872954823832E + 01$

$x(8) = -0.2828455330694E + 01$

$f(x^*) = 0.$

(98) *Quadratic 2 Variables* (Jamil and Yang 2013) [QUAD-2V]

$$f(x) = -3803.84 - 138.08x_1 - 232.92x_2 + 128.08x_1^2 + 203.64x_2^2 + 182.25x_1x_2.$$

$n = 2$. Initial point: $x_0 = [10, \quad 10]$. $f(x_0) = 43883.16$.

Solution: $x^* = [0.193877, \quad 0.485137]$, $f(x^*) = -3873.724182$.

(99) *Rump* (Jamil and Yang 2013) [RUMP]

$$f(x) = \left(333.75 - x_1^2 \right)x_2^6 + x_1^2 \left(11x_1^2x_2^2 - 121x_2^4 - 2 \right) + 5.5x_2^8 + \frac{x_1^2}{2x_2}.$$

$n = 2$. Initial point: $x_0 = [0.1, \quad 0.1]$. $f(x_0) = 0.480223$. Solution: $x^* = [0, 0]$, $f(x^*) = -68.500768$.

(100) *Extended Cliff* (Andrei 2008) [EX-CLIFF]

$$f(x) = \sum_{i=1}^{n/2} \left(\frac{x_{2i-1} - 3}{100}\right)^2 - (x_{2i-1} - x_{2i}) + \exp\left(20(x_{2i-1} - x_{2i})\right).$$

$n = 4$. Initial point: $x_0 = [1.5, \ldots 1.5]$. $f(x_0) = 2.0$.
   Solution:
   $x(1) = 0.2997569024910E + 01$ $x(3) = 0.2999807036205E + 01$
   $x(2) = 0.3147351602425E + 01$ $x(4) = 0.3149590021904E + 01$

$f(x^*) = 0.3995732282$

(101) *NONDIA (CUTE)* (Gould et al. 2013; Andrei 2020, p. 461) [NONDIA]

$$f(x) = (x_1 - 1)^2 + \sum_{i=2}^{n} 100\left(x_1 - x_{i-1}^2\right)^2.$$

$n = 10$. Initial point: $x_0 = [-1, \ldots, -1]$. $f(x_0) = 3604.0$.
   Solution:
   $x(1) = 0.1021254376141E - 01$ $x(6) = -0.1010653810733E + 00$
   $x(2) = -0.1010885682414E + 00$ $x(7) = 0.1010643028271E + 00$
   $x(3) = 0.1011218914368E + 00$ $x(8) = 0.1010268301976E + 00$
   $x(4) = -0.1011270448547E + 00$ $x(9) = -0.1010935500177E + 00$
   $x(5) = -0.1010903827437E + 00$ $x(10) = 0.1188971019331E - 01$

$f(x^*) = 0.989896.$

(102) *EG2 (SIN) (CUTE)* (Gould et al. 2013; Andrei 2020, p. 462) [EG2]

$$f(x) = \sum_{i=1}^{n-1} \sin\left(x_1 + x_i^2 - 1\right) + \frac{1}{2} \sin\left(x_n^2\right).$$

$n = 4$. Initial point: $x_0 = [1, \ldots, 1]$. $f(x_0) = 2.945148$.
   Solution:
   $x(1) = 0.8164615455672E + 01$ $x(3) = 0.3146890096240E + 02$
   $x(2) = 0.2539125727897E + 02$ $x(4) = 0.8872012206227E + 02$

$f(x^*) = -3.5.$

(103) *LIARWHD (CUTE)* (Gould et al. 2013; Andrei 2020, p. 465) [LIARWHD]

$$f(x) = \sum_{i=1}^{n} 4\left(x_i^2 - x_1\right)^2 + (x_i - 1)^2.$$

$n = 8$. Initial point: $x_0 = [4, \ldots, 4]$. $f(x_0) = 4680.0$. Solution: $x^* = [1, 1, 1, 1, 1, 1, 1, 1]$, $f(x^*) = 0$.

(104) *Full Hessian* (Andrei 2020, p. 459) [FULL-HES]

$$f(x) = \sum_{i=1}^{m} \left(\sum_{j=1}^{n} ijx_j^2 - 1\right)^2, \quad m = 50,$$

$n = 4$. Initial point: $x_0 = [1/n, 2/n, \ldots, n/n]$. $f(x_0) = 1660870.3125$.
Solution:
$$x(1) = 0.6607609302112E - 01 \qquad x(3) = 0.6802001878929E - 01$$
$$x(2) = -0.7560475528168E - 01 \quad x(4) = -0.2473350533439E - 02$$

$$f(x^*) = 12.1287128.$$

(105) *A Nonlinear Algebraic System* (Andrei 2013, pp. 49–51) [NALSYS]

$$f(x) = \left(x_2 \exp\left(-x_1^2\right) - 0.1\right)^2 + \left(x_2 - \sin\left(x_1\right)\right)^2.$$

$n = 2$. Initial point: $x_0 = [-2, \ 1]$. $f(x_0) = 0.0148992$.
Solution: $x^* = [-1.5169560, \ 0.998549]$, $f(x^*) = 0$.

(106) *ENGVAL8 (CUTE)* (Gould et al. 2013; Andrei 2020, p. 465) [ENGVAL8]

$$f(x) = \sum_{i=1}^{n-1} \left(x_i^2 + x_{i+1}^2\right)^2 - (7 - 8x_i).$$

$n = 4$. Initial point: $x_0 = [2, \ldots, 2]$. $f(x_0) = 219.0$.
Solution:
$$x(1) = -0.1145719143677E + 01 \quad x(3) = -0.9279944572218E + 00$$
$$x(2) = -0.6579579722912E + 00 \quad x(4) = -0.2796021711672E - 04$$

$$f(x^*) = -37.39004994.$$

(107) *DIXMAANA (CUTE)* (Gould et al. 2013; Andrei 2020, p. 464) [DIXMAANA]

$$f(x) = 1 + \sum_{i=1}^{n} \alpha x_i^2 \left(\frac{i}{n}\right)^{k1} + \sum_{i=1}^{n-1} \beta x_i^2 \left(x_{i+1} + x_{i+1}^2\right)^2 \left(\frac{i}{n}\right)^{k2}$$
$$+ \sum_{i=1}^{2m} \gamma x_i^2 x_{i+m}^4 \left(\frac{i}{n}\right)^{k3} + \sum_{i=1}^{m} \delta x_i x_{i+2m} \left(\frac{i}{n}\right)^{k4},$$

$m = n/4$, $x_0 = [2., 2., \ldots, 2.]$, $\alpha = 1$, $\beta = 0$, $\gamma = 0.125$, $\delta = 0.125$, $k1 = 0$, $k2 = 0$, $k3 = 0$, $k4 = 0$. $n = 10$. Initial point: $x_0 = [2, \ldots, 2]$. $f(x_0) = 74.0$.

Solution:

$x(1) = -0.2407049418505E - 04$ $x(6) = -0.3801287693504E - 04$
$x(2) = 0.1681285312734E - 04$ $x(7) = 0.1495632327834E - 04$
$x(3) = 0.5513302392268E - 04$ $x(8) = -0.2904469092107E - 04$
$x(4) = -0.7687904222039E - 05$ $x(9) = 0.5921868290259E - 05$
$x(5) = 0.1647687470278E - 04$ $x(10) = -0.7512179404971E - 04$

$f(x^*) = 1$.

(108) *DIXMAANB (CUTE)* (Gould et al. 2013; Andrei 2020, p. 464) [DIXMAANB]

$$f(x) = 1 + \sum_{i=1}^{n} \alpha x_i^2 \left(\frac{i}{n}\right)^{k1} + \sum_{i=1}^{n-1} \beta x_i^2 \left(x_{i+1} + x_{i+1}^2\right)^2 \left(\frac{i}{n}\right)^{k2}$$
$$+ \sum_{i=1}^{2m} \gamma x_i^2 x_{i+m}^4 \left(\frac{i}{n}\right)^{k3} + \sum_{i=1}^{m} \delta x_i x_{i+2m} \left(\frac{i}{n}\right)^{k4},$$

$m = n/4$, $x_0 = [2., 2., \ldots, 2.]$, $\alpha = 1$, $\beta = 0.0625$, $\gamma = 0.0625$, $\delta = 0.0625$, $k1 = 0$, $k2 = 0$, $k3 = 0$, $k4 = 1$. $n = 10$. Initial point: $x_0 = [2, \ldots, 2]$. $f(x_0) = 138.075$.

Solution:

$x(1) = -0.2268294044024E - 04$ $x(6) = 0.3366115293229E - 04$
$x(2) = -0.3895826103691E - 04$ $x(7) = -0.3344607567615E - 04$
$x(3) = 0.2923532181003E - 04$ $x(8) = 0.3561036384515E - 05$
$x(4) = 0.1052990442709E - 04$ $x(9) = -0.6044036129062E - 04$
$x(5) = 0.3144468647446E - 04$ $x(10) = -0.3911118508566E - 04$

$f(x^*) = 1$.

(109) *DIXMAANC (CUTE)* (Gould et al. 2013; Andrei 2020, p. 464) [DIXMAANC]

$$f(x) = 1 + \sum_{i=1}^{n} \alpha x_i^2 \left(\frac{i}{n}\right)^{k1} + \sum_{i=1}^{n-1} \beta x_i^2 \left(x_{i+1} + x_{i+1}^2\right)^2 \left(\frac{i}{n}\right)^{k2}$$
$$+ \sum_{i=1}^{2m} \gamma x_i^2 x_{i+m}^4 \left(\frac{i}{n}\right)^{k3} + \sum_{i=1}^{m} \delta x_i x_{i+2m} \left(\frac{i}{n}\right)^{k4},$$

$m = n/4$, $x_0 = [2., 2., \ldots, 2.]$, $\alpha = 1$, $\beta = 0.125$, $\gamma = 0.125$, $\delta = 0.125$, $k1 = 0$, $k2 = 0$, $k3 = 0$, $k4 = 0$. $n = 5$. Initial point: $x_0 = [2, \ldots, 2]$. $f(x_0) = 109.5$.
Solution:

$x(1) = 0.4654112359640E - 04$	$x(4) = -0.1118893687560E - 04$
$x(2) = -0.1254817539302E - 04$	$x(5) = 0.4202261312468E - 06$
$x(3) = 0.1419158777153E - 04$	

$f(x^*) = 1.$

(110) *DIAG-AUP1:* [DIAGAUP1]

$$f(x) = \sum_{i=1}^{n} \left(x_i^2 - x_1\right)^2 + \left(x_i^2 - 1\right)^2$$

$n = 5$. Initial point: $x_0 = [4, \ldots, 4]$. $f(x_0) = 4005.0$. Solution: $x^* = [1, 1, 1, 1, 1]$, $f(x^*) = 0$.

(111) *EG3 (COS) (CUTE)* (Gould et al. 2013; Andrei 2020, p. 462) [EG3-COS]

$$f(x) = \frac{1}{2} \cos\left(x_n^2\right) + \sum_{i=1}^{n-1} \cos\left(x_1 + x_i^2 - 1\right).$$

$n = 10$. Initial point: $x_0 = [2, \ldots, 2]$. $f(x_0) = 2.226137$.
Solution:

$x(1) = 0.4933618184739E + 01$	$x(6) = 0.2343287457984E + 01$
$x(2) = 0.5533914358914E + 01$	$x(7) = 0.4249394090899E + 01$
$x(3) = 0.5533922472445E + 01$	$x(8) = 0.5533883800774E + 01$
$x(4) = 0.4933623663556E + 01$	$x(9) = 0.2343311512128E + 01$
$x(5) = 0.4933618907462E + 01$	$x(10) = 0.4689537530417E + 01$

$f(x^*) = -9.449999.$

(112) *VARDIM* (Gould et al. 2013; Andrei 2020, p. 465) [VARDIM]

$$f(x) = \sum_{i=1}^{n} (x_i - 1)^2 + \left(\sum_{i=1}^{n} i x_i - \frac{n(n+1)}{2} \right)^2 + \left(\sum_{i=1}^{n} i x_i - \frac{n(n+1)}{2} \right)^4.$$

$n = 10$. Initial point: $x_0^i = 1 - \frac{i}{n}$, $i = 1, \ldots, n$. $f(x_0) = 2198553.1458$.
  Solution:
  $x(1) = 0.1000051790760E + 01$ $x(6) = 0.1000076913999E + 01$
  $x(2) = 0.9999979019194E + 00$ $x(7) = 0.1000017684842E + 01$
  $x(3) = 0.1000057409184E + 01$ $x(8) = 0.9999356124213E + 00$
  $x(4) = 0.1000024867354E + 01$ $x(9) = 0.9999594271384E + 00$
  $x(5) = 0.9999785818737E + 00$ $x(10) = 0.1000007117846E + 01$

$f(x^*) = 0$.

(113) *A Narrow Positive Cone* (Frimannslund and Steihaug 2011) [NARRCONE]

$$f(x) = \sum_{i=1}^{n/2} (9x_{2i-1} - x_{2i})(11x_{2i-1} - x_{2i}) + \frac{1}{2}x_{2i-1}^4,$$

$n = 2$. Initial point: $x_0 = [2, \ldots 2]$. $f(x_0) = 328.0$. Solution: $x^* = [0.996229, 9.961589]$, $f(x^*) = -0.499971$.
  (114) *Ackley Function* (Jamil and Yang 2013) [ACKLEY]

$$f(x) = -20 \exp\left(-0.2\sqrt{t_1/t_2} \right) - \exp\left(t_2/t_3 \right) + 20,$$

where

$$t_1 = \sum_{i=1}^{n} x_i^2, \quad t_2 = \sum_{i=1}^{n} \cos\left(2\pi x_i \right), \quad t_3 = n.$$

$n = 10$. Initial point: $x_0 = [2, \ldots 2]$. $f(x_0) = 3.875317$.
  Solution:
  $x(1) = -0.2343396213531E - 04$ $x(6) = 0.7616685348236E - 05$
  $x(2) = -0.3568660439254E - 04$ $x(7) = -0.4831649446963E - 04$
  $x(3) = -0.4195757893939E - 04$ $x(8) = -0.5669293726872E - 04$
  $x(4) = -0.7151680833778E - 04$ $x(9) = -0.2872996349272E - 04$
  $x(5) = 0.9954211011807E - 05$ $x(10) = 0.2281271967522E - 04$

$f(x^*) = -2.718123$.

(115) *Modified Wolfe Function* (Frimannslund and Steihaug 2011) [WOLFEmod]

$$f(x) = \sum_{i=1}^{n/2} \frac{1}{3} x_{2i-1}^3 + \frac{1}{2} x_{2i}^2 - \frac{2}{3} \left(\min \{x_{2i-1}, -1\} + 1 \right)^3.$$

$n = 2$. Initial point: $x_0 = [1, \ldots 1]$. $f(x_0) = 0.833333$. Solution: $x^* = [-3.414204, 0.118940E - 04]$, $f(x^*) = -3.885618$.

(116) *Peaks Function*: [PEAKS]

$$f(x) = 3(1 - x_1)^2 \exp\left(-x_1^2 - (x_2 + 1)^2\right)$$
$$-10\left(x_1/5 - x_1^3 - x_2^5\right) \exp\left(-x_1^2 - x_2^2\right)$$
$$- \exp\left(-(x_1 + 1) - x_2^2\right)/3.$$

$n = 2$. Initial point: $x_0 = [3, -2]$. $f(x_0) = 0.0004161$. Solution: $x^* = [0.228681, -1.625834]$, $f(x^*) = -6.547921$.

(117) *Function U18*: [U18]

$$f(x) = (x_1 + 5)^2 + (x_2 + 8)^2 + (x_3 + 7)^2 + 2x_1^2 x_2^2 + 4x_1^2 x_3^2.$$

$n = 3$. Initial point: $x_0 = [1, 1, 1]$. $f(x_0) = 187.0$.
  Solution:
$$x(1) = -0.1540925170299E - 01$$
$$x(2) = -0.7996172133856E + 01$$
$$x(3) = -0.6993305717752E + 01$$

$$f(x^*) = 24.923018.$$

(118) *Function U23*: [U23]

$$f(x) = x_1^3 \exp\left(x_2 - x_1^2 - 10(x_1 - x_2^2)\right).$$

$n = 2$. Initial point: $x_0 = [2, 2]$. $f(x_0) = 1.082682$. Solution: $x^* = [-1.652961, -0.572105]$, $f(x^*) = 0$.

(119) *Sum of Squares and Quartics*: [SUMSQUAR]

$$f(x) = \sum_{i=1}^{n} i\left(x_i^2 + x_i^4\right).$$

$n = 5$. Initial point: $x_0 = [1, \ldots 1]$. $f(x_0) = 30.0$. Solution: $x^* = [0, 0, 0, 0, 0]$, $f(x^*) = 0.0$.

(120) *Modified VARDIM:* [VARDIM8]

$$f(x) = \sum_{i=1}^{n} \left(x_i^2 - 1\right)^2 + \left(\sum_{i=1}^{n} ix_i - \frac{n(n+1)}{2}\right)^2 + \left(\sum_{i=1}^{n} ix_i - \frac{n(n+1)}{2}\right)^4$$
$$+ \left(\sum_{i=1}^{n} ix_i - \frac{n(n+1)}{2}\right)^6 + \left(\sum_{i=1}^{n} ix_i - \frac{n(n+1)}{2}\right)^8.$$

$n = 10$. Initial point: $x_0^i = 1 - \frac{i}{n}$, $i = 1, \ldots, n$. $f(x_0) = 0.483035E + 13$. Solution: $x^* = [1, \ldots, 1]$, $f(x^*) = 0$.

(121) *Module Function:* [MODULE]

$$f(x) = (|x_1| + |x_2|) \exp\left(-x_1^2 - x_2^2\right).$$

$n = 2$. Initial point: $x_0 = [1, 1]$, $f(x_0) = 0.270670$. Solution: $x^* = [3.071245, 2.968479]$, $f(x^*) = 0$.

(122) *PExp Function:* [PEXP]

$$f(x) = (x_1 - x_2)^2 \exp\left(-x_1^2 - x_2^2\right).$$

$n = 2$. Initial point: $x_0 = [1, -2]$, $f(x_0) = 0.0606415$. Solution: $x^* = [-0.035889, -0.035888]$, $f(x^*) = 0$.

(123) *Combination of EXP:* [COMB-EXP]

$$f(x) = \exp\left(-(x_1 - 4)^2 - (x_2 - 4)^2\right) + \exp\left(-(x_1 + 4)^2 - (x_2 - 4)^2\right)$$
$$+ 2\exp\left(-x_1^2 - x_2^2\right) + 2\exp\left(-x_1^2 - (x_2 + 4)^2\right).$$

$n = 2$. Initial point: $x_0 = [1, 1]$, $f(x_0) = 0.270670$. Solution: $x^* = [3.848219, -1.692257]$, $f(x^*) = 0$.

(124) *A Quadratic Function:* [QUADR1]

$$f(x) = 10x_1^2 - 5x_1x_2 + 10(x_2 - 3)^2.$$

$n = 2$. Initial point: $x_0 = [10, 10]$, $f(x_0) = 1990$. Solution: $x^* = [-0.8, 3.19999]$, $f(x^*) = -6.0$.

(125) *A Quadratic Function (1, 1):* [QUADR2]

$$f(x) = 5x_1^2 + x_2^2 + 4x_1x_2 - 14x_1 - 6x_2 + 20.$$

$n = 2$. Initial point: $x_0 = [0, 10]$, $f(x_0) = 60$. Solution: $x^* = [1, 1]$, $f(x^*) = 10$.

(126) *Sum of EXP:* [SUM-EXP]

$$f(x) = \sum_{i=1}^{n-1} \exp\left(x_i^2 + 2x_{i+1}^2\right).$$

$n = 50$. Initial point: $x_0 = [1, \ldots, 1]$, $f(x_0) = 984.191309$. Solution: $x^* = [0, \ldots, 0]$, $f(x^*) = 49$.

(127) *Camel Function* (Jamil and Yang 2013) [CAMEL]

$$f(x) = 2x_1^2 - 1.05x_1^4 + \frac{1}{6}x_1^6 + x_1 x_2^3.$$

$n = 2$. Initial point: $x_0 = [1, 1]$, $f(x_0) = 2.11666$. Solution: $x^* = [-0.0890547, 47.513875]$, $f(x^*) = -9552.512351$.

(128) *Sum and Prod of Modules Function:* [SP-MOD]

$$f(x) = \sum_{i=1}^{n} |x_i| + \prod_{i=1}^{n} |x_i|.$$

$n = 10$. Initial point: $x_0 = [1, \ldots, 1]$, $f(x_0) = 11$.

Solution:

$x(1) = -0.3971127419088E - 04$ $x(6) = 0.8316097004328E - 05$

$x(2) = 0.2566021146299E - 04$ $x(7) = 0.2292906350921E - 04$

$x(3) = 0.2219988860772E - 05$ $x(8) = -0.6682570806599E - 04$

$x(4) = -0.1007771028492E - 04$ $x(9) = -0.4341432742472E - 04$

$x(5) = -0.4452348098360E - 04$ $x(10) = 0.2980058191391E - 04$

$f(x^*) = 0.293478443E - 03.$

(129) *Trecanni Function* (Jamil and Yang 2013) [TRECANNI]

$$f(x) = x_1^4 + 4x_1^3 + 4x_1^2 + x_2^2.$$

$n = 2$. Initial point: $x_0 = [1, 1]$, $f(x_0) = 10$. Solution: $x^* = [0, 0]$, $f(x^*) = 0$.

(130) *Prod Variable and Module (Variable):* [ProV-MV]

$$f(x) = \sum_{i=1}^{n} x_i |x_i| + (x_i - i)^2.$$

$n = 10$. Initial point: $x_0 = [1, \ldots, 1]$, $f(x_0) = 295$.

Solution:

$x(1) = 0.1171401203889E + 01$ $x(6) = 0.2958293678336E + 01$

$x(2) = 0.2357897442087E + 01$ $x(7) = 0.3363568634896E + 01$

$x(3) = 0.2385642308096E + 01$ $x(8) = 0.3585153684128E + 01$

$x(4) = 0.2951388224001E + 01$ $x(9) = 0.4812009879818E + 01$

$x(5) = 0.3197515549144E + 01$ $x(10) = 0.4890643976612E + 01$

$f(x^*) = 202.044908.$

(131) *Scaled Quadratic Function:* [SCALQUAD]

$$f(x) = \sum_{i=1}^{n} i(x_i - 1) + i(x_i - 2)^2.$$

$n = 10$. Initial point: $x_0 = [1, \ldots, 1]$. $f(x_0) = 55$.
  Solution:
  $x(1) = 0.1228354359908E + 01$ $x(6) = 0.1686702271161E + 01$
  $x(2) = 0.1648985969190E + 01$ $x(7) = 0.1518358585878E + 01$
  $x(3) = 0.1767639832524E + 01$ $x(8) = 0.1422333928581E + 01$
  $x(4) = 0.1750649827990E + 01$ $x(9) = 0.1646114245137E + 01$
  $x(5) = 0.1596625198120E + 01$ $x(10) = 0.1486923164284E + 01$

$f(x^*) = 42.334677.$

(132) *Brasov Function:* [BRASOV]

$$f(x) = \sum_{i=1}^{n-1} (|x_i| + 2|x_{i+1}| - 9)^2 + (2|x_i| + |x_{i+1}| - 13)^2.$$

$n = 10$. Initial point: $x_0 = [1, \ldots, 1]$. $f(x_0) = 1224$.
  Solution:
  $x(1) = 0.4335899479442E + 01$ $x(6) = -0.3604128768683E + 01$
  $x(2) = 0.3330067280519E + 01$ $x(7) = 0.3760537815256E + 01$
  $x(3) = 0.3838850497961E + 01$ $x(8) = 0.3494501727656E + 01$
  $x(4) = 0.3572698821525E + 01$ $x(9) = -0.4003233955858E + 01$
  $x(5) = 0.3729292633436E + 01$ $x(10) = 0.2997353862478E + 01$

$f(x^*) = 69.322896.$

(133) *Multi PROD-SUM Function:* [PROD-SUM]

$$f(x) = |x_1| + \sum_{i=2}^{n} \left(|x_i| \sum_{j=1}^{i} |x_j| \right).$$

$n = 10$. Initial point: $x_0 = [1, \ldots, 1]$. $f(x_0) = 55$. Solution: $x^* = [0, \ldots, 0]$, $f(x^*) = 0$.

(134) *Multi PROD-PROD Function:* [PRODPROD]

$$f(x) = |x_1| + \sum_{i=2}^{n} \left(|x_i| \prod_{j=1}^{i} |x_j| \right).$$

$n = 10$. Initial point: $x_0 = [10, \ldots, 10]$. $f(x_0) = 0.1111E + 12$.

Solution:

$x(1) = -0.1063362156310E - 05$ $x(6) = -0.1039764689733E + 01$

$x(2) = 0.6739654650088E + 00$ $x(7) = -0.4805770813306E + 00$

$x(3) = -0.6580520005883E + 00$ $x(8) = 0.3226297302958E + 00$

$x(4) = 0.1285587626601E + 00$ $x(9) = 0.9317621908231E + 00$

$x(5) = -0.8536980202912E + 00$ $x(10) = 0.1107095794397E + 01$

$f(x^*) = 0.$

(135) *Multi PROD-SUM (cos):* [PS-COS]

$$f(x) = |x_1| \cos(x_1) + \sum_{i=2}^{n} \left(|x_i| \sum_{j=1}^{i} \cos(x_j) \right).$$

$n = 10$. Initial point: $x_0 = [10, \ldots, 10]$. $f(x_0) = -461.48934$.

Solution:

$x(1) = 0.3150821943752E + 01$ $x(6) = 0.1578895456040E + 02$

$x(2) = 0.9443720678143E + 01$ $x(7) = 0.1582814085482E + 02$

$x(3) = 0.3172734763737E + 01$ $x(8) = 0.1589674884209E + 02$

$x(4) = 0.9467772675187E + 01$ $x(9) = 0.1604993064465E + 02$

$x(5) = 0.9484630129254E + 01$ $x(10) = 0.1052232972564E + 02$

$f(x^*) = -690.239222.$

(136) *Multi PROD-PROD (cos) function:* [PP-COS]

$$f(x) = |x_1| \cos(x_1) + \sum_{i=2}^{n} \left(|x_i| \prod_{j=1}^{i} \cos(x_j) \right).$$

$n = 10$. Initial point: $x_0 = [10, \ldots, 10]$. $f(x_0) = -3.773267$.

Solution:

$x(1) = 0.9439893026895E + 01$ $x(6) = 0.9393126523959E + 01$

$x(2) = 0.1258400284984E + 02$ $x(7) = 0.9449118940688E + 01$

$x(3) = 0.1258900772732E + 02$ $x(8) = 0.1259804260487E + 02$

$x(4) = 0.9393102581192E + 01$ $x(9) = 0.1261901735516E + 02$

$x(5) = 0.9449123034167E + 01$ $x(10) = 0.6437292710163E + 01$

$f(x^*) = -66.177191.$

(137) *Multi PROD-SUM (sin) Function:* [PS-SIN]

$$f(x) = |x_1| \sin(x_1) + \sum_{i=2}^{n} \left(|x_i| \sum_{j=1}^{i} \sin(x_j) \right).$$

$n = 10$. Initial point: $x_0 = [10, \ldots, 10]$. $f(x_0) = -299.21161$.
Solution:

$x(1) = 0.1100449034905E + 02$ $x(6) = 0.1110123843662E + 02$
$x(2) = 0.1101536321108E + 02$ $x(7) = 0.1114895864822E + 02$
$x(3) = 0.1102894761787E + 02$ $x(8) = 0.1122792910065E + 02$
$x(4) = 0.1104626469194E + 02$ $x(9) = 0.1138633659904E + 02$
$x(5) = 0.1106935485532E + 02$ $x(10) = 0.1191579297819E + 02$

$f(x^*) = -613.165736$.

(138) *Multi PROD-PROD (sin) Function:* [PP-SIN]

$$f(x) = |x_1| \sin(x_1) + \sum_{i=2}^{n} \left(|x_i| \prod_{j=1}^{i} \sin(x_j) \right).$$

$n = 10$. Initial point: $x_0 = [10, \ldots, 10]$. $f(x_0) = -3.515404$.
Solution:

$x(1) = 0.1100995870993E + 02$ $x(6) = 0.7881575394324E + 01$
$x(2) = 0.4695258380142E + 01$ $x(7) = 0.4677132721497E + 01$
$x(3) = 0.1101144092912E + 02$ $x(8) = 0.1102575552025E + 02$
$x(4) = 0.7873235862038E + 01$ $x(9) = 0.1418235701698E + 02$
$x(5) = 0.7876686400784E + 01$ $x(10) = 0.7978776786912E + 01$

$f(x^*) = -69.291578$.

(139) *Bohachevsky Function* (Jamil and Yang 2013) [BOHA]

$$f(x) = x_1^2 + 2x_2^2 - 0.3 \cos(3\pi x_1) - 0.4 \cos(4\pi x_2) + 0.7.$$

$n = 2$. Initial point: $x_0 = [1, 1]$. $f(x_0) = 3.6$. Solution: $x^* = [0, 0]$, $f(x^*) = 0$.

(140) *Deckkers-Aarts Function* (Jamil and Yang 2013) [DECK-AAR]

$$f(x) = 10000x_1^2 + x_2^2 - \left(x_1^2 + x_2^2\right)^2 + 0.000001 \left(x_1^2 + x_2^2\right)^4.$$

$n = 2$. Initial point: $x_0 = [1, 1]$. $f(x_0) = 9997$. Solution: $x^* = [0, 26.586773]$, $f(x^*) = -249293.0182$.

References

Andrei, N., (2003). *Modele, Probleme de Test şi Aplicaţii de Programare Matematică [Models, Test problems and Applications of Mathematical Programming]*, Editura Tehnică [Technical Press], Bucharest, Romania.

Andrei, N., (2008). An unconstrained optimization test functions collection. Advanced Modeling and Optimization, vol.10, No.1, pp.147–161.

Andrei, N., (2013). *Nonlinear Optimization Applications using the GAMS Technology.* Springer Optimization and Its Applications, vol.81. Springer Science + Business Media, New York.

Andrei, N., (2020). *Nonlinear Conjugate Gradient Methods for Unconstrained Optimization.* Springer Optimization and Its Applications, vol.158. Springer + Business Media, New York, USA.

Brent, R.P., (1973). Chapter 4: An Algorithm with Guaranteed Convergence for Finding a Zero of a Function. In: *Algorithms for Minimization without Derivatives,* Englewood Cliffs, NJ: Prentice-Hall.

Floudas, C.A., Pardalos, M.P., Adjiman, C.S., Esposito, W.R., Gümüs, Z.H., Harding, S. T., Klepeis, J.L., Meyer, C.A., & Schweiger, C.A., (1999). *Handbook of Test Problems in Local and Global Optimization,* Kluwer Academic Publishers, Dordrecht.

Frimannslund, L., & Steihaug, T., (2011). On a new method for derivative free optimization, International Journal on Advances in Software, 4(3-4), 244–255.

Gould, N.I.M., Orban, D., & Toint, Ph.L., (2013). *CUTEst: A Constrained and Unconstrained Testing Environment with safe threads.* Technical Report, RAL-TR-2013–005, May 2013.

Himmelblau, D.M., (1972). *Applied Nonlinear Programming.* McGraw-Hill Book Company, New York.

Jamil, M., & Yang, X-S., (2013). A literature survey of benchmark functions for global optimization problems, Int. Journal of Mathematical Modelling and Numerical Optimisation, Vol. 4, No. 2, pp. 150–194.

Moré, J.J., Garbow, B.S., & Hillstrom, K.E., (1981). Testing unconstrained optimization software. ACM Transaction on Mathematical Software, vol.7, no.1, pp.17–41.

Schittkowski, K., (1987). *More test examples for nonlinear programming codes.* Springer Verlag, Berlin.

Annex C

Detailed Results for 30 Large-Scale Problems

This annex includes the results of the optimization process of DEEPS for solving 30 unconstrained optimization problems. In the table below, N is the number of trial points, M is the number of local trial points, n is the number of variables, *iter* is the number of iterations to get the minimum point, *nfunc* is the number of function evaluations, *time* is the CPU computing time in centiseconds, *vfomin* is the minimum value of the minimizing function, *vf0* is the function value in the initial point, *lobnd*, *upbnd*, *lobndc*, and *upbndc* are the bounds of the domains at the start of the optimization (Initial) and at the end of the optimization (Final), *BR*, *BCR*, *CR*, and *LS* are the number of bounds *lobnd* and *upbnd* reduction, the number of the local bounds *lobndc* and *upbndc* reduction, the number of the complex reduction and the number of line searches, respectively. If $LS = 0$, then the line search procedure from step 12 is inhibited.

```
                 A Two Level Random Search Method for Unconstrained Optimization
                 ===================================================================
                                Total results for 30 large-scale problems
          27. DENSCHNA
            n=100      N= 500     M= 100      iter=    54      nfunc=   2726250        time=   2813
                         lobnd              upbnd              lobndc             upbndc
Initial  -0.5000000000000E+02    0.5000000000000E+02  -0.2000000000000E+01    0.2000000000000E+01
Final    -0.3125000000000E+01    0.3125000000000E+01  -0.2441406250000E-03    0.2441406250000E-03
     BR=   4  BCR=  13  CR=  44  LS=   0  vfomin= 0.1933041622350E-05  vf0= 0.3976246221006E+03
-----------------------------------------------------------------------------------------------------

          28. DENSCHNB
            n=200      N= 800     M=  10      iter=   475      nfunc=   4178799        time=   3365
                         lobnd              upbnd              lobndc             upbndc
Initial  -0.5000000000000E+03    0.5000000000000E+03  -0.2000000000000E+01    0.2000000000000E+01
Final    -0.3906250000000E+01    0.3906250000000E+01  -0.2441406250000E-03    0.2441406250000E-03
     BR=   7  BCR=  13  CR= 444  LS= 238  vfomin= 0.7607056471717E-05  vf0= 0.7129000000000E+04
-----------------------------------------------------------------------------------------------------

          33. MATYAS
            n=200      N=1000     M=  50      iter=    39      nfunc=   1977715        time=   1582
                         lobnd              upbnd              lobndc             upbndc
Initial  -0.1000000000000E+02    0.1000000000000E+02  -0.1000000000000E+01    0.1000000000000E+01
Final    -0.1000000000000E+02    0.1000000000000E+02  -0.9765625000000E-03    0.9765625000000E-03
     BR=   0  BCR=  10  CR=  38  LS=  15  vfomin= 0.1906786878984E-05  vf0= 0.4027200000000E+01
-----------------------------------------------------------------------------------------------------

          44. ELIPSOID
            n=100      N= 100     M=  50      iter=   594      nfunc=   3025219        time=   1352
                         lobnd              upbnd              lobndc             upbndc
Initial  -0.5000000000000E+01    0.5000000000000E+01  -0.2000000000000E+01    0.2000000000000E+01
Final    -0.6250000000000E+00    0.6250000000000E+00  -0.4882812500000E-03    0.4882812500000E-03
     BR=   3  BCR=  12  CR= 364  LS= 274  vfomin= 0.8927972405540E-06  vf0= 0.3383500000000E+06
-----------------------------------------------------------------------------------------------------

          63. BROYDEN1
            n=200      N=  50     M=  10      iter=  1184      nfunc=    700626        time=   1069
                         lobnd              upbnd              lobndc             upbndc
Initial  -0.2000000000000E+01    0.2000000000000E+01  -0.1000000000000E+01    0.1000000000000E+01
Final    -0.2000000000000E+01    0.2000000000000E+01  -0.3814697265625E-05    0.3814697265625E-05
     BR=   0  BCR=  18  CR=   1  LS=   0  vfomin= 0.1869593904842E-03  vf0= 0.2110000000000E+03
-----------------------------------------------------------------------------------------------------

          69. TRIG-1
            n=200      N= 100     M=  50      iter=    10      nfunc=     52095        time=    399
                         lobnd              upbnd              lobndc             upbndc
Initial  -0.2000000000000E+01    0.2000000000000E+01  -0.1000000000000E+01    0.1000000000000E+01
Final    -0.5000000000000E+00    0.5000000000000E+00  -0.1000000000000E+01    0.1000000000000E+01
     BR=   2  BCR=   0  CR=   6  LS=   0  vfomin= 0.3927344805542E-05  vf0= 0.6972774389540E+04
-----------------------------------------------------------------------------------------------------

          72. BROWN
            n=200      N=  50     M=  50      iter=    14      nfunc=     36127        time=    296
                         lobnd              upbnd              lobndc             upbndc
Initial  -0.1000000000000E+01    0.4000000000000E+01  -0.1500000000000E+01    0.1500000000000E+01
Final    -0.1000000000000E+01    0.4000000000000E+01  -0.7500000000000E+00    0.7500000000000E+00
     BR=   0  BCR=   1  CR=  13  LS=   0  vfomin= 0.9136497330827E-05  vf0= 0.3800870492365E+01
-----------------------------------------------------------------------------------------------------

          75. COSMIN
            n=200      N=   5     M=  10      iter=   498      nfunc=     28611        time=     38
                         lobnd              upbnd              lobndc             upbndc
Initial  -0.5000000000000E+03    0.5000000000000E+03  -0.1000000000000E+02    0.1000000000000E+02
Final    -0.6250000000000E+02    0.6250000000000E+02  -0.1525878906250E-03    0.1525878906250E-03
     BR=   3  BCR=  16  CR= 443  LS=   0  vfomin= 0.6234069231056E-05  vf0= 0.6500000000000E+04
-----------------------------------------------------------------------------------------------------

          78. ENGVAL1
            n=200      N= 100     M=  50      iter=    56      nfunc=    285028        time=    245
                         lobnd              upbnd              lobndc             upbndc
Initial  -0.5000000000000E+03    0.5000000000000E+03  -0.5000000000000E+02    0.5000000000000E+02
Final    -0.3906250000000E+01    0.3906250000000E+01  -0.9765625000000E-01    0.9765625000000E-01
     BR=   7  BCR=   9  CR=  37  LS=   0  vfomin= 0.2460315490512E+03  vf0= 0.1174100000000E+05
-----------------------------------------------------------------------------------------------------

          79. EX-PEN
            n=500      N=  10     M= 100      iter=   380      nfunc=    386099        time=    810
                         lobnd              upbnd              lobndc             upbndc
Initial  -0.5000000000000E+03    0.5000000000000E+03  -0.2000000000000E+03    0.2000000000000E+03
Final    -0.9765625000000E+00    0.9765625000000E+00  -0.9536743164062E-04    0.9536743164062E-04
     BR=   9  BCR=  21  CR= 231  LS=   0  vfomin= 0.4266196551499E+03  vf0= 0.1746550388459E+16
-----------------------------------------------------------------------------------------------------
```

```
80. BROYDENP
  n=500     N= 100      M= 100         iter=   1179       nfunc=   12010364            time=  41069
                           lobnd                upbnd             lobndc                 upbndc
Initial  -0.5000000000000E+01    0.5000000000000E+01  -0.2000000000000E+01    0.2000000000000E+01
Final    -0.5000000000000E+01    0.5000000000000E+01  -0.7629394531250E-05    0.7629394531250E-05
  BR=   0  BCR=  18  CR=1167  LS=   0  vfomin= 0.7896764624658E-03   vf0= 0.5320000000000E+03
------------------------------------------------------------------------------------------------

85. NONDQUAR
  n=500     N= 100      M=  50         iter=     13       nfunc=     67719            time=     90
                           lobnd                upbnd             lobndc                 upbndc
Initial  -0.5000000000000E+02    0.5000000000000E+02  -0.1000000000000E+02    0.1000000000000E+02
Final    -0.1562500000000E+01    0.1562500000000E+01  -0.1000000000000E+02    0.1000000000000E+02
  BR=   5  BCR=   0  CR=   7  LS=   0  vfomin= 0.3947652312005E-08   vf0= 0.4034200000000E+05
------------------------------------------------------------------------------------------------

86. ARWHEAD
  n=400     N=  50      M=  50         iter=    508       nfunc=   1316673            time=   2942
                           lobnd                upbnd             lobndc                 upbndc
Initial  -0.5000000000000E+02    0.5000000000000E+02  -0.1000000000000E+02    0.1000000000000E+02
Final    -0.3125000000000E+01    0.3125000000000E+01  -0.9536743164062E-05    0.9536743164062E-05
  BR=   4  BCR=  20  CR= 496  LS=   0  vfomin= 0.1520943530238E-03   vf0= 0.1197000000000E+04
------------------------------------------------------------------------------------------------

89. DENSCHNF
  n=500     N=   5      M=  10         iter=   2738       nfunc=    156891            time=    617
                           lobnd                upbnd             lobndc                 upbndc
Initial  -0.5000000000000E+01    0.5000000000000E+01  -0.5000000000000E+01    0.5000000000000E+01
Final    -0.6250000000000E+00    0.6250000000000E+00  -0.2384185791016E-05    0.2384185791016E-05
  BR=   3  BCR=  21  CR=2243  LS=   0  vfomin= 0.4679348483033E-02   vf0= 0.1040000000000E+06
------------------------------------------------------------------------------------------------

91. BORSEC6
  n=500     N= 100      M=  50         iter=     17       nfunc=     87802            time=    268
                           lobnd                upbnd             lobndc                 upbndc
Initial   0.0000000000000E+00    0.4000000000000E+00  -0.1000000000000E+01    0.1000000000000E+01
Final     0.0000000000000E+00    0.5000000000000E+00  -0.1000000000000E+01    0.1000000000000E+01
  BR=   3  BCR=   0  CR=  12  LS=   0  vfomin= 0.1515258658432E-03   vf0= 0.8274826754904E+26
------------------------------------------------------------------------------------------------

103. LIARWHD
  n=100     N=  50      M= 100         iter=    687       nfunc=   3499045            time=   1676
                           lobnd                upbnd             lobndc                 upbndc
Initial  -0.1000000000000E+03    0.1000000000000E+03  -0.1000000000000E+02    0.1000000000000E+02
Final    -0.1562500000000E+01    0.1562500000000E+01  -0.3814697265625E-04    0.3814697265625E-04
  BR=   6  BCR=  18  CR= 680  LS=   0  vfomin= 0.2427940007984E-04   vf0= 0.5850000000000E+05
------------------------------------------------------------------------------------------------

104. FULL-HES
  n=400     N=  10      M= 100         iter=     41       nfunc=     41802            time=    508
                           lobnd                upbnd             lobndc                 upbndc
Initial  -0.5000000000000E+03    0.5000000000000E+03  -0.1000000000000E+02    0.1000000000000E+02
Final    -0.9765625000000E+00    0.9765625000000E+00  -0.3051757812500E-03    0.3051757812500E-03
  BR=   9  BCR=  15  CR=  21  LS=   0  vfomin= 0.1212871290753E+02   vf0= 0.6936927728460E+14
------------------------------------------------------------------------------------------------

107. DIXMAANA
  n=500     N=   5      M=  10         iter=     28       nfunc=      1598            time=     24
                           lobnd                upbnd             lobndc                 upbndc
Initial   0.0000000000000E+00    0.1000000000000E+01  -0.1000000000000E+00    0.1000000000000E+00
Final     0.0000000000000E+00    0.1000000000000E+01  -0.1250000000000E+00    0.1250000000000E-01
  BR=   0  BCR=   3  CR=  27  LS=  14  vfomin= 0.1000032590334E+01   vf0= 0.4063500000000E+04
------------------------------------------------------------------------------------------------

108. DIXMAANB
  n=500     N= 100      M=  50         iter=     10       nfunc=     51691            time=   1003
                           lobnd                upbnd             lobndc                 upbndc
Initial   0.0000000000000E+00    0.1000000000000E+01  -0.1000000000000E+00    0.1000000000000E+00
Final     0.0000000000000E+00    0.1000000000000E+01  -0.1000000000000E+00    0.1000000000000E+00
  BR=   0  BCR=   0  CR=   9  LS=   1  vfomin= 0.1000030776510E+01   vf0= 0.7495937500000E+04
------------------------------------------------------------------------------------------------

109. DIXMAANC
  n=500     N= 100      M=  50         iter=     22       nfunc=    113066            time=   2723
                           lobnd                upbnd             lobndc                 upbndc
Initial   0.0000000000000E+00    0.1000000000000E+01  -0.1000000000000E+00    0.1000000000000E+00
Final     0.0000000000000E+00    0.1000000000000E+01  -0.9765625000000E-04    0.9765625000000E-04
  BR=   0  BCR=  10  CR=  21  LS=   0  vfomin= 0.1000029993149E+01   vf0= 0.1304550000000E+05
------------------------------------------------------------------------------------------------

110. DIAGAUP1
  n=100     N= 100      M=  50         iter=    241       nfunc=   1251550            time=    524
                           lobnd                upbnd             lobndc                 upbndc
Initial   0.0000000000000E+00    0.1000000000000E+05  -0.1000000000000E+03    0.1000000000000E+03
Final     0.0000000000000E+00    0.6103515625000E+00  -0.4768371582031E-04    0.4768371582031E-04
  BR=  14  BCR=  21  CR= 226  LS=   0  vfomin= 0.2597022323503E-04   vf0= 0.8010000000000E+05
------------------------------------------------------------------------------------------------
```

```
     111. EG3-COS
       n=500      N=   5     M=  10        iter=   3332      nfunc=     190828        time=    821
                             lobnd                  upbnd                lobndc                upbndc
Initial    0.0000000000000E+00   0.4000000000000E+01  -0.1000000000000E+00   0.1000000000000E+00
Final      0.0000000000000E+00   0.4000000000000E+01  -0.6103515625000E-05   0.6103515625000E-05
       BR=   0  BCR=  14  CR= 212  LS=   0  vfomin=-0.4994993713630E+03   vf0= 0.1412206087357E+03
-------------------------------------------------------------------------------------------------

     112. VARDIM
       n=100      N=  50     M= 100        iter=    301      nfunc=    1533231        time=    658
                             lobnd                  upbnd                lobndc                upbndc
Initial    0.0000000000000E+00   0.1000000000000E+02  -0.5000000000000E+00   0.5000000000000E+00
Final      0.0000000000000E+00   0.6250000000000E+00  -0.1220703125000E-03   0.1220703125000E-03
       BR=   4  BCR=  12  CR= 245  LS=   0  vfomin= 0.1333805427545E-04   vf0= 0.1310583696893E+15
-------------------------------------------------------------------------------------------------

     119. SUMSQUAR
       n=500      N= 100     M=1000        iter=     50      nfunc=    5006609        time=  18461
                             lobnd                  upbnd                lobndc                upbndc
Initial   -0.1000000000000E+02   0.1000000000000E+02  -0.1100000000000E+01   0.1100000000000E+01
Final     -0.6250000000000E-02   0.6250000000000E-02  -0.4296875000000E-02   0.4296875000000E-02
       BR=   4  BCR=   8  CR=   9  LS=  18  vfomin= 0.7079248254927E-02   vf0= 0.2505000000000E+06
-------------------------------------------------------------------------------------------------

     120. VARDIM8
       n=100      N=  50     M= 100        iter=    238      nfunc=    1212244        time=    523
                             lobnd                  upbnd                lobndc                upbndc
Initial    0.0000000000000E+00   0.1000000000000E+00  -0.1000000000000E+00   0.1000000000000E+00
Final      0.0000000000000E+00   0.5000000000000E+00  -0.1953125000000E-03   0.1953125000000E-03
       BR=   1  BCR=   9  CR= 217  LS=   0  vfomin= 0.5671097610006E-05   vf0= 0.1717629476525E+29
-------------------------------------------------------------------------------------------------

     126. SUM-EXP
       n=500      N=   5     M=  50        iter=     28      nfunc=       7193        time=     40
                             lobnd                  upbnd                lobndc                upbndc
Initial    0.1000000000000E+01   0.4000000000000E+01   0.1000000000000E+00   0.2000000000000E+01
Final      0.1250000000000E+00   0.5000000000000E+00   0.3906250000000E-03   0.7812500000000E-02
       BR=   3  BCR=   8  CR=  18  LS=   0  vfomin= 0.4990001347310E+03   vf0= 0.1002268292467E+05
-------------------------------------------------------------------------------------------------

     128. SP-MOD
       n=500      N=  50     M=  50        iter=     22      nfunc=      56795        time=    109
                             lobnd                  upbnd                lobndc                upbndc
Initial   -0.5000000000000E+01   0.5000000000000E+01   0.0000000000000E+00   0.1000000000000E+01
Final     -0.2500000000000E+01   0.2500000000000E+01   0.0000000000000E+00   0.1000000000000E+01
       BR=   1  BCR=   0  CR=  13  LS=   4  vfomin= 0.9444412279940E-01   vf0= 0.5010000000000E+03
-------------------------------------------------------------------------------------------------

     130. ProV-MV
       n=100      N=  10     M=  10        iter=   5001      nfunc=     550642        time=    222
                             lobnd                  upbnd                lobndc                upbndc
Initial   -0.5000000000000E+02   0.5000000000000E+02   0.0000000000000E+00   0.1000000000000E+01
Final     -0.7812500000000E+00   0.7812500000000E+00   0.0000000000000E+00   0.0000000000000E+00
       BR=   6  BCR=4933  CR=   0  LS=   0  vfomin= 0.1990001436291E+06   vf0= 0.3284500000000E+06
-------------------------------------------------------------------------------------------------

     132. BRASOV
```

```
Initial   -0.5000000000000E+01   0.5000000000000E+01  -0.1000000000000E+01   0.1000000000000E+01
Final     -0.5000000000000E+01   0.5000000000000E+01  -0.152878906250E-04    0.1525878906250E-04
       BR=   0  BCR=  16  CR=  53  LS=   0  vfomin= 0.1589333398727E+04   vf0= 0.2706400000000E+05
-------------------------------------------------------------------------------------------------

     133. PROD-SUM
       n=300      N=  50     M=  50        iter=     16      nfunc=      41334        time=    660
                             lobnd                  upbnd                lobndc                upbndc
Initial   -0.5000000000000E+01   0.5000000000000E+01  -0.1000000000000E+01   0.1000000000000E+01
Final     -0.2500000000000E+01   0.2500000000000E+01  -0.1000000000000E+01   0.1000000000000E+01
       BR=   1  BCR=   0  CR=   8  LS=   0  vfomin= 0.1519490671575E-02   vf0= 0.4515000000000E+05
-------------------------------------------------------------------------------------------------
```

Annex D

Detailed Results for 140 Problems

This annex includes the results of the optimization process of DEEPS for solving 140 unconstrained optimization problems. In the table below, N is the number of trial points, M is the number of local trial points, n is the number of variables, *iter* is the number of iterations to get the minimum point, *nfunc* is the number of function evaluations, *time* is the CPU computing time in centiseconds, *vfomin* is the minimum value of the minimizing function, *vf0* is the function value in the initial point, *lobnd*, *upbnd*, *lobndc*, and *upbndc* are the bounds of the domains at the start of the optimization (Initial) and at the end of the optimization (Final), *BR*, *BCR*, *CR*, and *LS* are the number of bounds *lobnd* and *upbnd* reduction, the number of the local bounds *lobndc* and *upbndc* reduction, the number of the complex reduction and the number of line searches, respectively. If $LS = 0$, then the line search procedure from step 12 is inhibited.

© The Author(s), under exclusive license to Springer Nature Switzerland AG 2021
N. Andrei, *A Derivative-free Two Level Random Search Method for Unconstrained Optimization*, SpringerBriefs in Optimization,
https://doi.org/10.1007/978-3-030-68517-1

A Two Level Random Search Method for Unconstrained Optimization
===
Total results for 140 problems

```
  1. WEBER-1
  n= 2      N=   2    M=    5       iter=   119      nfunc=       1485        time=     0
                      lobnd             upbnd             lobndc              upbndc
Initial  -0.1000000000000E+03   0.1000000000000E+03  -0.2000000000000E+01   0.2000000000000E+01
Final    -0.1000000000000E+03   0.1000000000000E+03  -0.4882812500000E-03   0.4882812500000E-03
  BR=   0  BCR=  12  CR=  13  LS=  14  vfomin=-0.2644531378352E+03   vf0=-0.3747313737140E+02
-------------------------------------------------------------------------------------------

  2. WEBER-2
  n= 2      N=   3    M=    5       iter=    67      nfunc=       1318        time=     0
                      lobnd             upbnd             lobndc              upbndc
Initial  -0.5000000000000E+03   0.5000000000000E+03  -0.1500000000000E+02   0.1500000000000E+02
Final    -0.5000000000000E+03   0.5000000000000E+03  -0.1430511474609E-04   0.1430511474609E-04
  BR=   0  BCR=  20  CR=  66  LS=  30  vfomin= 0.9560744054913E+01   vf0= 0.7859432489718E+02
-------------------------------------------------------------------------------------------

  3. WEBER-3
  n= 2      N=   5    M=   10       iter=   113      nfunc=       6198        time=     1
                      lobnd             upbnd             lobndc              upbndc
Initial  -0.5000000000000E+03   0.5000000000000E+03  -0.1000000000000E+03   0.1000000000000E+03
Final    -0.5000000000000E+03   0.5000000000000E+03  -0.9536743164062E-04   0.9536743164062E-04
  BR=   0  BCR=  20  CR= 109  LS=  80  vfomin= 0.8749847970082E+01   vf0= 0.7860286479337E+02
-------------------------------------------------------------------------------------------

  4. ENZIMES
  n= 4      N=  30    M=   50       iter=    28      nfunc=      42903        time=     2
                      lobnd             upbnd             lobndc              upbndc
Initial  -0.5000000000000E+01   0.5000000000000E+01  -0.2000000000000E+01   0.2000000000000E+01
Final    -0.5000000000000E+01   0.5000000000000E+01  -0.1250000000000E+00   0.1250000000000E+00
  BR=   0  BCR=   4  CR=  27  LS=  21  vfomin= 0.3087657632221E-03   vf0= 0.5313172272109E-02
-------------------------------------------------------------------------------------------

  5. REACTOR
  n= 6      N=  50    M=  100       iter=  1375      nfunc=    6989983        time=   197
                      lobnd             upbnd             lobndc              upbndc
Initial   0.0000000000000E+00   0.5000000000000E+01  -0.2000000000000E+01   0.2000000000000E+01
Final     0.0000000000000E+00   0.6250000000000E+00  -0.2441406250000E-03   0.2441406250000E-03
  BR=   3  BCR=  13  CR=   0  LS=1024  vfomin= 0.7472925581554E-04   vf0= 0.1961733675060E+08
-------------------------------------------------------------------------------------------

  6. ROBOT
  n= 8      N=   5    M=   10       iter=    43      nfunc=       2459        time=     0
                      lobnd             upbnd             lobndc              upbndc
Initial  -0.5000000000000E+01   0.5000000000000E+01  -0.1000000000000E+01   0.1000000000000E+01
Final    -0.5000000000000E+01   0.5000000000000E+01  -0.7812500000000E-02   0.7812500000000E-02
  BR=   0  BCR=   7  CR=  42  LS=  23  vfomin= 0.1828017526180E-07   vf0= 0.5334258881257E+01
-------------------------------------------------------------------------------------------

  7. SPECTR
  n= 4      N=   5    M=   10       iter=     4      nfunc=        233        time=     0
                      lobnd             upbnd             lobndc              upbndc
Initial   0.0000000000000E+00   0.2000000000000E+00   0.1000000000000E-01   0.5000000000000E+00
Final     0.0000000000000E+00   0.2000000000000E+00   0.5000000000000E-02   0.2500000000000E+00
  BR=   0  BCR=   1  CR=   3  LS=   2  vfomin= 0.8316382216967E+01   vf0= 0.9958700480657E+01
-------------------------------------------------------------------------------------------

  8. ESTIMP
  n= 4      N= 100    M=  500       iter=    54      nfunc=    2705727        time=    99
                      lobnd             upbnd             lobndc              upbndc
Initial  -0.9000000000000E+03   0.9000000000000E+03  -0.5000000000000E+03   0.5000000000000E+03
Final    -0.4500000000000E+03   0.4500000000000E+03  -0.9765625000000E+00   0.9765625000000E+00
  BR=   1  BCR=   9  CR=   0  LS=   7  vfomin= 0.3185724691657E-01   vf0= 0.2905300235663E+01
-------------------------------------------------------------------------------------------

  9. PROPAN
  n= 5      N=9900    M=  100       iter=    10      nfunc=   10013013        time=   274
                      lobnd             upbnd             lobndc              upbndc
Initial   0.0000000000000E+00   0.5000000000000E+03  -0.1000000000000E+02   0.1000000000000E+02
Final     0.0000000000000E+00   0.5000000000000E+03  -0.1000000000000E+02   0.1000000000000E+02
  BR=   0  BCR=   0  CR=   6  LS=   1  vfomin= 0.2467602087429E-04   vf0= 0.3312269269234E+08
-------------------------------------------------------------------------------------------

  10. GEAR-1
  n= 2      N=   5    M=    3       iter=    15      nfunc=        342        time=     0
                      lobnd             upbnd             lobndc              upbndc
Initial   0.0000000000000E+00   0.5000000000000E+01  -0.1000000000000E+01   0.1000000000000E+01
Final     0.0000000000000E+00   0.5000000000000E+01  -0.2500000000000E+00   0.2500000000000E+00
  BR=   0  BCR=   2  CR=  14  LS=  10  vfomin= 0.1744152005590E+01   vf0= 0.2563325000000E+04
-------------------------------------------------------------------------------------------
```

```
    11. HHD
    n=   8      N=  50      M= 300      iter=  2404      nfunc=  36265585        time=   1487
                           lobnd                 upbnd                lobndc                upbndc
Initial  -0.4000000000000E+01   0.4000000000000E+01  -0.1000000000000E+01   0.1000000000000E+01
Final    -0.4000000000000E+01   0.4000000000000E+01  -0.2441406250000E-03   0.2441406250000E-03
    BR=   0  BCR=  12  CR=2403  LS=2332  vfomin= 0.9966084682095E-04   vf0= 0.1905692550902E+00
----------------------------------------------------------------------------------------------

    12. NEURO
    n=   6      N=1000      M=   6      iter=  1044      nfunc=   7999853        time=    219
                           lobnd                 upbnd                lobndc                upbndc
Initial  -0.1000000000000E+01   0.1000000000000E+01  -0.1000000000000E+00   0.1000000000000E+00
Final    -0.1000000000000E+01   0.1000000000000E+01  -0.7812500000000E-03   0.7812500000000E-03
    BR=   0  BCR=   7  CR=1043  LS= 944  vfomin= 0.854501892614E-04   vf0= 0.2391760016000E+02
----------------------------------------------------------------------------------------------

    13. COMBUST
    n=  10      N=   3      M=  10      iter=    25      nfunc=     856         time=      0
                           lobnd                 upbnd                lobndc                upbndc
Initial  -0.1000000000000E+02   0.1000000000000E+02  -0.1000000000000E+00   0.1000000000000E+00
Final    -0.1000000000000E+02   0.1000000000000E+02  -0.1000000000000E+00   0.1000000000000E+00
    BR=   0  BCR=   0  CR=  24  LS=  16  vfomin= 0.4061987800161E-08   vf0= 0.1219988990749E+03
----------------------------------------------------------------------------------------------

    14. CIRCUIT
    n=   9      N=  50      M=  50      iter=  3731      nfunc=   9654682        time=   1244
                           lobnd                 upbnd                lobndc                upbndc
Initial  -0.1000000000000E+01   0.1000000000000E+01  -0.1000000000000E-01   0.1000000000000E-01
Final    -0.1000000000000E+01   0.1000000000000E+01  -0.9765625000000E-05   0.9765625000000E-05
    BR=   0  BCR=  10  CR=3730  LS=3456  vfomin= 0.1036184837525E-03   vf0= 0.2964578187893E+04
----------------------------------------------------------------------------------------------

    15. THERM
    n=   3      N= 100      M= 500      iter=    15      nfunc=    752179        time=    251
                           lobnd                 upbnd                lobndc                upbndc
Initial  -0.1000000000000E+01   0.1000000000000E+01  -0.1000000000000E+01   0.1000000000000E+01
Final    -0.5000000000000E+00   0.5000000000000E+00  -0.5000000000000E+00   0.5000000000000E+00
    BR=   1  BCR=   1  CR=   4  LS=   0  vfomin= 0.1742216236340E+03   vf0= 0.2335910048036E+10
----------------------------------------------------------------------------------------------

    16. GEAR-2
    n=   4      N=  10      M=  50      iter=     6      nfunc=     3090        time=      0
                           lobnd                 upbnd                lobndc                upbndc
Initial   0.0000000000000E+00   0.1000000000000E+01  -0.1000000000000E+00   0.1000000000000E+00
Final     0.0000000000000E+00   0.1000000000000E+01  -0.1000000000000E+00   0.1000000000000E+00
    BR=   0  BCR=   0  CR=   5  LS=   2  vfomin= 0.3886716443010E-13   vf0= 0.7370818569964E-03
----------------------------------------------------------------------------------------------

    17. BANANA
    n=   2      N=  10      M=  10      iter=    13      nfunc=     1452        time=      0
                           lobnd                 upbnd                lobndc                upbndc
Initial  -0.2000000000000E+01   0.2000000000000E+01  -0.5000000000000E+00   0.5000000000000E+00
Final    -0.2000000000000E+01   0.2000000000000E+01  -0.3125000000000E-01   0.3125000000000E-01
    BR=   0  BCR=   4  CR=  12  LS=   0  vfomin= 0.5223120133290E-08   vf0= 0.2420000000000E+02
----------------------------------------------------------------------------------------------

    18. FRE-ROTH
    n=   2      N=  10      M= 100      iter=    19      nfunc=    19219        time=      0
                           lobnd                 upbnd                lobndc                upbndc
Initial  -0.5000000000000E+03   0.5000000000000E+03  -0.2000000000000E+01   0.2000000000000E+01
Final    -0.3906250000000E+01   0.3906250000000E+01  -0.2500000000000E+01   0.2000000000000E+01
    BR=   7  BCR=   3  CR=  11  LS=   0  vfomin= 0.4898425368048E+02   vf0= 0.4005000000000E+03
----------------------------------------------------------------------------------------------

    19. WHI-HOL
    n=   2      N= 500      M= 100      iter=     8      nfunc=    409182        time=      1
                           lobnd                 upbnd                lobndc                upbndc
Initial  -0.5000000000000E+03   0.5000000000000E+03  -0.2000000000000E+01   0.2000000000000E+01
Final    -0.7812500000000E+01   0.7812500000000E+01  -0.2000000000000E+01   0.2000000000000E+01
    BR=   6  BCR=   0  CR=   1  LS=   0  vfomin= 0.9096890029946E-10   vf0= 0.1228198400000E+02
----------------------------------------------------------------------------------------------

    20. MI-CAN
    n=   4      N=  10      M=  50      iter=    11      nfunc=     5546        time=      1
                           lobnd                 upbnd                lobndc                upbndc
Initial  -0.5000000000000E+01   0.5000000000000E+01  -0.2000000000000E+01   0.2000000000000E+01
Final    -0.2500000000000E+01   0.2500000000000E+01  -0.1250000000000E+00   0.1250000000000E+00
    BR=   1  BCR=   4  CR=   0  LS=   0  vfomin= 0.5097821607941E-08   vf0= 0.1300897728708E+02
----------------------------------------------------------------------------------------------

    21. HIMM-1
    n=   2      N=  20      M= 100      iter=    17      nfunc=    34183        time=      1
                           lobnd                 upbnd                lobndc                upbndc
Initial  -0.5000000000000E+02   0.5000000000000E+02  -0.2000000000000E+01   0.2000000000000E+01
Final    -0.5000000000000E+02   0.5000000000000E+02  -0.1562500000000E-01   0.1562500000000E-01
    BR=   0  BCR=   7  CR=  16  LS=   0  vfomin= 0.3509702847612E-07   vf0= 0.2820895744000E+04
----------------------------------------------------------------------------------------------
```

```
    22. 3-CAMEL
    n=  2      N=  20     M= 100      iter=    6      nfunc=      12055       time=     0
                         lobnd                    upbnd                  lobndc                  upbndc
Initial  -0.5000000000000E+01   0.5000000000000E+01  -0.2000000000000E+01   0.2000000000000E+01
Final    -0.5000000000000E+01   0.5000000000000E+01  -0.1000000000000E+01   0.1000000000000E+01
  BR=   0  BCR=   1  CR=   5  LS=  0  vfomin=-0.1031628453412E+01   vf0= 0.3600768000000E+01
-----------------------------------------------------------------------------------------------

    23. 6-CAMEL
    n=  2      N=  20     M=   5      iter=   13      nfunc=       1448       time=     0
                         lobnd                    upbnd                  lobndc                  upbndc
Initial  -0.5000000000000E+01   0.5000000000000E+01  -0.2000000000000E+01   0.2000000000000E+01
Final    -0.5000000000000E+01   0.5000000000000E+01  -0.5000000000000E+00   0.5000000000000E+00
  BR=   0  BCR=   2  CR=  12  LS=  5  vfomin= 0.9837324056667E-10   vf0= 0.6700000000000E+01
-----------------------------------------------------------------------------------------------

    24. WOOD
    n=  4      N= 100     M=  50      iter=   41      nfunc=     210805       time=     4
                         lobnd                    upbnd                  lobndc                  upbndc
Initial  -0.5000000000000E+03   0.5000000000000E+03  -0.2000000000000E+01   0.2000000000000E+01
Final    -0.5000000000000E+03   0.5000000000000E+03  -0.3906250000000E-02   0.3906250000000E-02
  BR=   0  BCR=   9  CR=  39  LS=  0  vfomin= 0.3415656241886E-07   vf0= 0.1919200000000E+05
-----------------------------------------------------------------------------------------------

    25. QUADR-2
    n=  2      N= 100     M=   5      iter=   73      nfunc=      39075       time=     0
                         lobnd                    upbnd                  lobndc                  upbndc
Initial  -0.5000000000000E+05   0.5000000000000E+05  -0.2000000000000E+01   0.2000000000000E+01
Final    -0.5000000000000E+05   0.5000000000000E+05  -0.3125000000000E-01   0.3125000000000E-01
  BR=   0  BCR=   6  CR=  23  LS=  0  vfomin= 0.1100000000004E+02   vf0= 0.9850440000000E+04
-----------------------------------------------------------------------------------------------

    26. SHEKEL
    n=  2      N= 100     M=  50      iter=   25      nfunc=     126221       time=     2
                         lobnd                    upbnd                  lobndc                  upbndc
Initial  -0.5000000000000E+03   0.5000000000000E+03  -0.2000000000000E+01   0.2000000000000E+01
Final    -0.5000000000000E+03   0.5000000000000E+03  -0.1562500000000E-01   0.1562500000000E-01
  BR=   0  BCR=   7  CR=  24  LS=  0  vfomin=-0.1008600149530E+02   vf0=-0.4236176330972E+01
-----------------------------------------------------------------------------------------------

    27. DENSCHNA
    n=  8      N= 500     M= 100      iter=   29      nfunc=    1464284       time=   119
                         lobnd                    upbnd                  lobndc                  upbndc
Initial  -0.5000000000000E+02   0.5000000000000E+02  -0.2000000000000E+01   0.2000000000000E+01
Final    -0.3125000000000E+01   0.3125000000000E+01  -0.3906250000000E-02   0.3906250000000E-02
  BR=   4  BCR=   9  CR=  24  LS=  0  vfomin= 0.9678327750962E-08   vf0= 0.3180996976805E+02
-----------------------------------------------------------------------------------------------

    28. DENSCHNB
    n=  2      N= 800     M=  10      iter=   13      nfunc=     112880       time=     0
                         lobnd                    upbnd                  lobndc                  upbndc
Initial  -0.5000000000000E+03   0.5000000000000E+03  -0.2000000000000E+01   0.2000000000000E+01
Final    -0.7812500000000E+01   0.7812500000000E+01  -0.1000000000000E+01   0.1000000000000E+01
  BR=   6  BCR=   1  CR=   6  LS=  3  vfomin= 0.6599092624990E-11   vf0= 0.6585000000000E+04
-----------------------------------------------------------------------------------------------

    29. DENSCHNC
    n=  8      N= 100     M= 500      iter=  236      nfunc=   11823238       time=  1036
                         lobnd                    upbnd                  lobndc                  upbndc
Initial  -0.5000000000000E+01   0.5000000000000E+01  -0.2000000000000E+01   0.2000000000000E+01
Final    -0.2500000000000E+01   0.2500000000000E+01  -0.1953125000000E-02   0.1953125000000E-02
  BR=   1  BCR=  10  CR= 234  LS=213  vfomin= 0.2030922856474E-07   vf0= 0.5220203819919E+04
-----------------------------------------------------------------------------------------------

    30. GRIEWANK
    n=  2      N=1000     M=  50      iter=    3      nfunc=     151167       time=     3
                         lobnd                    upbnd                  lobndc                  upbndc
Initial  -0.5000000000000E+01   0.5000000000000E+01  -0.2000000000000E+01   0.2000000000000E+01
Final    -0.5000000000000E+01   0.5000000000000E+01  -0.1000000000000E+01   0.1000000000000E+01
  BR=   0  BCR=   1  CR=   2  LS=  0  vfomin= 0.2505775587025E-09   vf0= 0.7251294749507E+00
-----------------------------------------------------------------------------------------------

    31. BRENT
    n=  2      N=  10     M=  50      iter=   19      nfunc=       9774       time=     0
                         lobnd                    upbnd                  lobndc                  upbndc
Initial  -0.5130000000000E+03   0.5130000000000E+03  -0.1000000000000E+00   0.1000000000000E+00
Final    -0.2565000000000E+03   0.2565000000000E+03  -0.1000000000000E+00   0.1000000000000E+00
  BR=   1  BCR=   0  CR=  17  LS=  0  vfomin= 0.6794175748481E-10   vf0= 0.1990000000000E+03
-----------------------------------------------------------------------------------------------

    32. BOOTH
    n=  2      N=1000     M=  50      iter=    4      nfunc=     201408       time=     2
                         lobnd                    upbnd                  lobndc                  upbndc
Initial  -0.1000000000000E+02   0.1000000000000E+02  -0.2000000000000E+01   0.2000000000000E+01
Final    -0.1000000000000E+02   0.1000000000000E+02  -0.2000000000000E+01   0.2000000000000E+01
  BR=   0  BCR=   0  CR=   3  LS=  1  vfomin= 0.2147910448310E-09   vf0= 0.1700000000000E+02
-----------------------------------------------------------------------------------------------
```

```
33. MATYAS
  n= 2      N=1000     M= 50        iter=    3      nfunc=    151310           time=    2
                       lobnd                  upbnd                  lobndc                 upbndc
Initial  -0.1000000000000E+02   0.1000000000000E+02  -0.1000000000000E+01   0.1000000000000E+01
Final    -0.1000000000000E+02   0.1000000000000E+02  -0.1000000000000E+01   0.1000000000000E+01
  BR=   0  BCR=   0  CR=   2  LS=   1  vfomin= 0.5443086223290E-10   vf0= 0.1072000000000E+00
------------------------------------------------------------------------------------------------

34. COLVILLE
  n= 3      N= 10      M= 500       iter=   20      nfunc=    100171           time=    2
                       lobnd                  upbnd                  lobndc                 upbndc
Initial  -0.1000000000000E+02   0.1000000000000E+02  -0.1000000000000E+01   0.1000000000000E+01
Final    -0.6250000000000E+00   0.6250000000000E+00  -0.3125000000000E-01   0.3125000000000E-01
  BR=   4  BCR=   5  CR=  15  LS=   0  vfomin= 0.1930369016075E-07   vf0= 0.9615280400000E+06
------------------------------------------------------------------------------------------------

35. EASOM
  n= 2      N=1000     M= 500       iter=    6      nfunc=   3002322           time=  122
                       lobnd                  upbnd                  lobndc                 upbndc
Initial  -0.1000000000000E+03   0.1000000000000E+03  -0.1000000000000E+01   0.1000000000000E+01
Final    -0.1000000000000E+03   0.1000000000000E+03  -0.5000000000000E+00   0.5000000000000E+00
  BR=   0  BCR=   1  CR=   5  LS=   0  vfomin=-0.9999999999144E+00   vf0=-0.3030892310248E-04
------------------------------------------------------------------------------------------------

36. BEALE
  n= 8      N= 10      M= 10        iter=  181      nfunc=     21192           time=    1
                       lobnd                  upbnd                  lobndc                 upbndc
Initial  -0.2000000000000E+01   0.2000000000000E+01  -0.1000000000000E+01   0.1000000000000E+01
Final    -0.2000000000000E+01   0.2000000000000E+01  -0.4882812500000E-03   0.4882812500000E-03
  BR=   0  BCR=  11  CR= 180  LS=   0  vfomin= 0.2962368797323E-04   vf0= 0.3931547600000E+02
------------------------------------------------------------------------------------------------

37. POWELL
  n= 4      N=1000     M= 500       iter=    5      nfunc=   2502034           time=   50
                       lobnd                  upbnd                  lobndc                 upbndc
Initial  -0.2000000000000E+01   0.2000000000000E+01  -0.1000000000000E+00   0.1000000000000E+00
Final    -0.2000000000000E+01   0.2000000000000E+01  -0.1000000000000E+00   0.1000000000000E+00
  BR=   0  BCR=   0  CR=   4  LS=   0  vfomin= 0.7775209572113E-09   vf0= 0.2150000000000E+03
------------------------------------------------------------------------------------------------

38. McCORM
  n= 2      N=1000     M= 500       iter=    8      nfunc=   4010395           time=   62
                       lobnd                  upbnd                  lobndc                 upbndc
Initial  -0.1500000000000E+01   0.4000000000000E+01  -0.1000000000000E-01   0.1000000000000E-01
Final    -0.1500000000000E+01   0.4000000000000E+01  -0.1000000000000E-01   0.1000000000000E-01
  BR=   0  BCR=   0  CR=   7  LS=   0  vfomin=-0.1913222954963E+01   vf0= 0.7090702573174E+01
------------------------------------------------------------------------------------------------

39. HIMM-2
  n= 2      N= 200     M= 100       iter=   10      nfunc=    200879           time=    2
                       lobnd                  upbnd                  lobndc                 upbndc
Initial  -0.2000000000000E+01   0.2000000000000E+01  -0.1000000000000E+00   0.1000000000000E+00
Final    -0.2000000000000E+01   0.2000000000000E+01  -0.2500000000000E-01   0.2500000000000E-01
  BR=   0  BCR=   2  CR=   9  LS=   0  vfomin= 0.4820864500816E-09   vf0= 0.1060000000000E+03
------------------------------------------------------------------------------------------------

40. LEON
  n= 2      N=1000     M= 50        iter=    6      nfunc=    304418           time=    2
                       lobnd                  upbnd                  lobndc                 upbndc
Initial  -0.1000000000000E+02   0.1000000000000E+02  -0.1000000000000E+01   0.1000000000000E+01
Final    -0.2500000000000E+01   0.2500000000000E+01  -0.5000000000000E+00   0.5000000000000E+00
  BR=   2  BCR=   1  CR=   3  LS=   0  vfomin= 0.1359029423164E-08   vf0= 0.7490384000000E+03
------------------------------------------------------------------------------------------------

41. PRICE4
  n= 2      N= 500     M= 50        iter=    4      nfunc=    101964           time=    0
                       lobnd                  upbnd                  lobndc                 upbndc
Initial  -0.1000000000000E+02   0.1000000000000E+02  -0.1000000000000E+01   0.1000000000000E+01
Final    -0.2500000000000E+01   0.2500000000000E+01  -0.1000000000000E+01   0.1000000000000E+01
  BR=   2  BCR=   0  CR=   1  LS=   2  vfomin= 0.1946503233539E-11   vf0= 0.3307193600000E+02
------------------------------------------------------------------------------------------------

42. ZETTL
  n= 2      N= 500     M= 50        iter=    6      nfunc=    151312           time=    2
                       lobnd                  upbnd                  lobndc                 upbndc
Initial  -0.1000000000000E+02   0.1000000000000E+02  -0.1000000000000E+01   0.1000000000000E+01
Final    -0.1000000000000E+02   0.1000000000000E+02  -0.1000000000000E+01   0.1000000000000E+01
  BR=   0  BCR=   0  CR=   5  LS=   3  vfomin=-0.3791237214486E-02   vf0= 0.2500000000000E+00
------------------------------------------------------------------------------------------------

43. SPHERE
  n= 8      N=1000     M= 50        iter=   23      nfunc=   1171077           time=   38
                       lobnd                  upbnd                  lobndc                 upbndc
Initial  -0.5000000000000E+01   0.5000000000000E+01  -0.2000000000000E+01   0.2000000000000E+01
Final    -0.5000000000000E+01   0.5000000000000E+01  -0.7812500000000E-02   0.7812500000000E-02
  BR=   0  BCR=   8  CR=  22  LS=   0  vfomin= 0.5509141299242E-08   vf0= 0.6480000000000E+03
------------------------------------------------------------------------------------------------
```

```
    44. ELIPSOID
    n=  8      N=1000      M=  50      iter=    21      nfunc=   1069087      time=    35
                           lobnd               upbnd               lobndc              upbndc
Initial   -0.5000000000000E+01   0.5000000000000E+01  -0.2000000000000E+01   0.2000000000000E+01
Final     -0.5000000000000E+01   0.5000000000000E+01  -0.7812500000000E-02   0.7812500000000E-02
    BR=   0  BCR=   8  CR=  20  LS=   9  vfomin= 0.386705346080E-08   vf0= 0.2040000000000E+03
------------------------------------------------------------------------------------------------

    45. HIMM-3
    n=  2      N=2000      M=  50      iter=     4      nfunc=    414840      time=     2
                           lobnd               upbnd               lobndc              upbndc
Initial   -0.2000000000000E+01   0.2000000000000E+01  -0.1000000000000E+01   0.1000000000000E+01
Final     -0.5000000000000E+00   0.5000000000000E+00  -0.1000000000000E+01   0.1000000000000E+01
    BR=   2  BCR=   0  CR=   1  LS=   0  vfomin= 0.5922563192900E+01   vf0= 0.3330769000000E+07
------------------------------------------------------------------------------------------------

    46. HIMM-4
    n=  3      N=2000      M=  50      iter=     3      nfunc=    302066      time=     4
                           lobnd               upbnd               lobndc              upbndc
Initial   -0.2000000000000E+01   0.2000000000000E+01  -0.1000000000000E+01   0.1000000000000E+01
Final     -0.2000000000000E+01   0.2000000000000E+01  -0.1000000000000E+01   0.1000000000000E+01
    BR=   0  BCR=   0  CR=   2  LS=   0  vfomin= 0.2120841536218E-08   vf0= 0.8400000000000E+01
------------------------------------------------------------------------------------------------

    47. HIMM-5
    n=  2      N=  40      M=  50      iter=     5      nfunc=     10250      time=     0
                           lobnd               upbnd               lobndc              upbndc
Initial    0.1000000000000E+01   0.2000000000000E+01  -0.5000000000000E-01   0.5000000000000E-01
Final      0.1000000000000E+01   0.2000000000000E+01  -0.5000000000000E-01   0.5000000000000E-01
    BR=   0  BCR=   0  CR=   4  LS=   0  vfomin= 0.5247703287149E-10   vf0= 0.4598493014643E+00
------------------------------------------------------------------------------------------------

    48. ZIRILLI
    n=  2      N=1000      M= 500      iter=     6      nfunc=   3012421      time=    15
                           lobnd               upbnd               lobndc              upbndc
Initial   -0.2000000000000E+03   0.2000000000000E+03  -0.1000000000000E-02   0.1000000000000E-02
Final     -0.2500000000000E+02   0.2500000000000E+02  -0.1000000000000E-02   0.1000000000000E-02
    BR=   3  BCR=   0  CR=   2  LS=   0  vfomin=-0.3523860737867E+00   vf0= 0.8000000000000E+01
------------------------------------------------------------------------------------------------

    49. STYBLIN
    n=  2      N= 100      M=  50      iter=     3      nfunc=     15519      time=     0
                           lobnd               upbnd               lobndc              upbndc
Initial   -0.5000000000000E+01   0.5000000000000E+01  -0.1000000000000E+00   0.1000000000000E+00
Final     -0.5000000000000E+01   0.5000000000000E+01  -0.1000000000000E+00   0.1000000000000E+00
    BR=   0  BCR=   0  CR=   2  LS=   0  vfomin=-0.7833233140751E+02   vf0=-0.2000000000000E+02
------------------------------------------------------------------------------------------------

    50. TRID
    n=  2      N= 100      M=  50      iter=     3      nfunc=     15480      time=     0
                           lobnd               upbnd               lobndc              upbndc
Initial   -0.4000000000000E+01   0.4000000000000E+01  -0.1000000000000E+01   0.1000000000000E+01
Final     -0.4000000000000E+01   0.4000000000000E+01  -0.1000000000000E+01   0.1000000000000E+01
    BR=   0  BCR=   0  CR=   2  LS=   0  vfomin=-0.1999999999988E+01   vf0= 0.7000000000000E+01
------------------------------------------------------------------------------------------------

    51. SCALQ
    n=  2      N= 100      M= 500      iter=    28      nfunc=   1404941      time=    12
                           lobnd               upbnd               lobndc              upbndc
Initial   -0.5000000000000E+01   0.5000000000000E+01  -0.1000000000000E+01   0.1000000000000E+01
Final     -0.6250000000000E+00   0.6250000000000E+00  -0.1000000000000E+01   0.1000000000000E+01
    BR=   3  BCR=   0  CR=  16  LS=  16  vfomin= 0.9942813112235E-08   vf0= 0.8100000016000E+06
------------------------------------------------------------------------------------------------

    52. SCHIT-1
    n=  3      N= 200      M= 100      iter=    21      nfunc=    425993      time=     6
                           lobnd               upbnd               lobndc              upbndc
Initial    0.0000000000000E+00   0.5000000000000E+02  -0.1000000000000E+02   0.1000000000000E+02
Final      0.0000000000000E+00   0.1250000000000E+02  -0.7812500000000E-01   0.7812500000000E-01
    BR=   2  BCR=   7  CR=  18  LS=   0  vfomin= 0.8592871044861E-08   vf0= 0.6290000000000E+03
------------------------------------------------------------------------------------------------

    53. SCHIT-2
    n=  6      N= 200      M= 100      iter=    25      nfunc=    508583      time=    13
                           lobnd               upbnd               lobndc              upbndc
Initial    0.0000000000000E+00   0.5000000000000E+02  -0.1000000000000E+02   0.1000000000000E+02
Final      0.0000000000000E+00   0.5000000000000E+02  -0.7812500000000E-01   0.7812500000000E-01
    BR=   0  BCR=   7  CR=  24  LS=   0  vfomin= 0.3292756425536E-07   vf0= 0.7500000000000E+02
------------------------------------------------------------------------------------------------

    54. SCHIT-3
    n=  2      N=  20      M= 100      iter=     2      nfunc=      4070      time=     0
                           lobnd               upbnd               lobndc              upbndc
Initial    0.0000000000000E+00   0.5000000000000E+01  -0.1000000000000E+01   0.1000000000000E+01
Final      0.0000000000000E+00   0.5000000000000E+01  -0.1000000000000E+01   0.1000000000000E+01
    BR=   0  BCR=   0  CR=   1  LS=   0  vfomin= 0.7731990565053E+00   vf0= 0.8768604814560E+02
------------------------------------------------------------------------------------------------
```

```
     55. BROWN-A
     n= 5     N= 20     M= 50        iter=    4       nfunc=      4139          time=      0
                         lobnd                   upbnd                 lobndc                    upbndc
Initial  -0.2000000000000E+01   0.2000000000000E+01  -0.1000000000000E+01   0.1000000000000E+01
Final    -0.2000000000000E+01   0.2000000000000E+01  -0.1000000000000E+01   0.1000000000000E+01
     BR=  0  BCR=  0  CR=  3  LS=  0  vfomin= 0.254192808954E-08    vf0= 0.3693847656250E+02
-------------------------------------------------------------------------------------------

     56. KELLEY
     n= 4     N= 20     M= 500       iter=    6       nfunc=     60194          time=      1
                         lobnd                   upbnd                 lobndc                    upbndc
Initial  -0.2000000000000E+01   0.2000000000000E+01  -0.1000000000000E+01   0.1000000000000E+01
Final    -0.2000000000000E+01   0.2000000000000E+01  -0.1000000000000E+01   0.1000000000000E+01
     BR=  0  BCR=  0  CR=  5  LS=  1  vfomin= 0.5918576884523E-09   vf0= 0.4400000000000E+02
-------------------------------------------------------------------------------------------

     57. NONSYS
     n= 4     N= 20     M= 500       iter=   18       nfunc=    180402          time=      4
                         lobnd                   upbnd                 lobndc                    upbndc
Initial  -0.2000000000000E+01   0.2000000000000E+01  -0.1000000000000E+01   0.1000000000000E+01
Final    -0.2000000000000E+01   0.2000000000000E+01  -0.3125000000000E-01   0.3125000000000E-01
     BR=  0  BCR=  5  CR=  0  LS=  5  vfomin= 0.2281055628062E-08   vf0= 0.1890000000000E+03
-------------------------------------------------------------------------------------------

     58. ZANGWILL
     n= 2     N= 100    M= 50        iter=    2       nfunc=     10363          time=      0
                         lobnd                   upbnd                 lobndc                    upbndc
Initial  -0.3000000000000E+01   0.3000000000000E+01  -0.1000000000000E+01   0.1000000000000E+01
Final    -0.3000000000000E+01   0.3000000000000E+01  -0.1000000000000E+01   0.1000000000000E+01
     BR=  0  BCR=  0  CR=  1  LS=  0  vfomin=-0.1819999999993E+02   vf0=-0.1660000000000E+02
-------------------------------------------------------------------------------------------

     59. CIRCULAR
     n= 3     N= 100    M= 50        iter=    2       nfunc=     10392          time=      1
                         lobnd                   upbnd                 lobndc                    upbndc
Initial  -0.2000000000000E+01   0.2000000000000E+01  -0.1000000000000E+01   0.1000000000000E+01
Final    -0.2000000000000E+01   0.2000000000000E+01  -0.1000000000000E+01   0.1000000000000E+01
     BR=  0  BCR=  0  CR=  1  LS=  0  vfomin=-0.3923048451913E+00   vf0= 0.1000000000000E+02
-------------------------------------------------------------------------------------------

     60. POLEXP
     n= 2     N= 100    M= 50        iter=   20       nfunc=    102703          time=      3
                         lobnd                   upbnd                 lobndc                    upbndc
Initial  -0.1000000000000E+01   0.1000000000000E+01  -0.5000000000000E+00   0.5000000000000E+00
Final    -0.1000000000000E+01   0.1000000000000E+01  -0.5000000000000E+00   0.5000000000000E+00
     BR=  0  BCR=  0  CR= 19  LS=  0  vfomin= 0.4610244291330E-06   vf0= 0.1082682265893E+01
-------------------------------------------------------------------------------------------

     61. DULCE
     n= 2     N= 100    M= 50        iter=    1       nfunc=      5200          time=      0
                         lobnd                   upbnd                 lobndc                    upbndc
Initial  -0.2000000000000E+02   0.2000000000000E+02  -0.1000000000000E+02   0.1000000000000E+02
Final    -0.2000000000000E+02   0.2000000000000E+02  -0.1000000000000E+02   0.1000000000000E+02
     BR=  0  BCR=  0  CR=  0  LS=  0  vfomin=-0.5000000000000E+00   vf0= 0.4000000000000E+00
-------------------------------------------------------------------------------------------

     62. CRAGLEVY
     n= 4     N=1000    M= 50        iter=    3       nfunc=    154209          time=     12
                         lobnd                   upbnd                 lobndc                    upbndc
Initial  -0.2000000000000E+01   0.2000000000000E+01  -0.1000000000000E+01   0.1000000000000E+01
Final    -0.2000000000000E+01   0.2000000000000E+01  -0.1000000000000E+01   0.1000000000000E+01
     BR=  0  BCR=  0  CR=  2  LS=  0  vfomin= 0.5069703944580E-12   vf0= 0.2266182511289E+01
-------------------------------------------------------------------------------------------

     63. BROYDEN1
     n= 5     N= 50     M= 10        iter=   42       nfunc=     23205          time=      1
                         lobnd                   upbnd                 lobndc                    upbndc
Initial  -0.2000000000000E+01   0.2000000000000E+01  -0.1000000000000E+01   0.1000000000000E+01
Final    -0.2000000000000E+01   0.2000000000000E+01  -0.2441406250000E-03   0.2441406250000E-03
     BR=  0  BCR= 12  CR=  0  LS=  0  vfomin= 0.3674886987132E-07   vf0= 0.1600000000000E+02
-------------------------------------------------------------------------------------------

     64. BROYDEN2
     n= 8     N= 100    M= 500       iter=   40       nfunc=   2005756          time=     84
                         lobnd                   upbnd                 lobndc                    upbndc
Initial  -0.2000000000000E+01   0.2000000000000E+01  -0.1000000000000E+01   0.1000000000000E+01
Final    -0.2000000000000E+01   0.2000000000000E+01  -0.9765625000000E-03   0.9765625000000E-03
     BR=  0  BCR= 10  CR= 39  LS=  0  vfomin= 0.9392975052918E+00   vf0= 0.1900000000000E+02
-------------------------------------------------------------------------------------------

     65. BROYDEN3
     n= 50    N= 100    M= 50        iter=  126       nfunc=    654057          time=    150
                         lobnd                   upbnd                 lobndc                    upbndc
Initial  -0.2000000000000E+01   0.2000000000000E+01  -0.1000000000000E+01   0.1000000000000E+01
Final    -0.2000000000000E+01   0.2000000000000E+01  -0.3051757812500E-04   0.3051757812500E-04
     BR=  0  BCR= 15  CR=125  LS=  0  vfomin= 0.9960497839120E-05   vf0= 0.6100000000000E+02
-------------------------------------------------------------------------------------------
```

```
66. FULRANK1
   n= 5      N= 100     M=  50       iter=    12     nfunc=      62010        time=      1
                          lobnd              upbnd              lobndc             upbndc
Initial  -0.3000000000000E+01   0.3000000000000E+01  -0.1000000000000E+01   0.1000000000000E+01
Final    -0.3000000000000E+01   0.3000000000000E+01  -0.1250000000000E+00   0.1250000000000E+00
   BR=   0  BCR=   3  CR=  11  LS=   0  vfomin= 0.6698885009208E-11  vf0= 0.4856000000000E+04
-----------------------------------------------------------------------------------------------

67. FULRANK2
   n= 8      N=  10     M= 500       iter=    16     nfunc=      80264        time=      5
                          lobnd              upbnd              lobndc             upbndc
Initial  -0.3000000000000E+01   0.3000000000000E+01  -0.1000000000000E+01   0.1000000000000E+01
Final    -0.3000000000000E+01   0.3000000000000E+01  -0.1250000000000E+00   0.1250000000000E+00
   BR=   0  BCR=   3  CR=  15  LS=   2  vfomin= 0.2290198352507E-05  vf0= 0.3846000000000E+05
-----------------------------------------------------------------------------------------------

68. FULRANK3
   n= 10     N=  10     M= 500       iter=    22     nfunc=     110320        time=      8
                          lobnd              upbnd              lobndc             upbndc
Initial  -0.3000000000000E+01   0.3000000000000E+01  -0.1000000000000E+01   0.1000000000000E+01
Final    -0.3000000000000E+01   0.3000000000000E+01  -0.3125000000000E-01   0.3125000000000E-01
   BR=   0  BCR=   5  CR=  21  LS=   0  vfomin= 0.3706326698194E-05  vf0= 0.1076320000000E+06
-----------------------------------------------------------------------------------------------

69. TRIG-1
   n= 5      N= 100     M=  50       iter=     7     nfunc=      36160        time=      5
                          lobnd              upbnd              lobndc             upbndc
Initial  -0.2000000000000E+01   0.2000000000000E+01  -0.1000000000000E+01   0.1000000000000E+01
Final    -0.2000000000000E+01   0.2000000000000E+01  -0.1000000000000E+01   0.1000000000000E+01
   BR=   0  BCR=   0  CR=   6  LS=   0  vfomin= 0.2037153258108E-08  vf0= 0.1165737899047E-01
-----------------------------------------------------------------------------------------------

70. TRIG-2
   n= 9      N= 100     M= 500       iter=    12     nfunc=     601804        time=    154
                          lobnd              upbnd              lobndc             upbndc
Initial  -0.1000000000000E+02   0.1000000000000E+02  -0.5000000000000E+00   0.5000000000000E+00
Final    -0.1000000000000E+02   0.1000000000000E+02  -0.1250000000000E+00   0.1250000000000E+00
   BR=   0  BCR=   2  CR=  11  LS=   0  vfomin= 0.1026965671220E-07  vf0= 0.8201586330328E-01
-----------------------------------------------------------------------------------------------

71. TRIG-3
   n= 20     N= 100     M=1000       iter=    10     nfunc=    1001534        time=    568
                          lobnd              upbnd              lobndc             upbndc
Initial  -0.2000000000000E+02   0.2000000000000E+02  -0.1500000000000E+01   0.1500000000000E+01
Final    -0.2000000000000E+02   0.2000000000000E+02  -0.3750000000000E+00   0.3750000000000E+00
   BR=   0  BCR=   2  CR=   9  LS=   0  vfomin= 0.4625184409097E-07  vf0= 0.3614762514618E+01
-----------------------------------------------------------------------------------------------

72. BROWN
   n= 10     N=  50     M=  50       iter=    10     nfunc=      25898        time=      5
                          lobnd              upbnd              lobndc             upbndc
Initial  -0.1000000000000E+01   0.4000000000000E+01  -0.1500000000000E+01   0.1500000000000E+01
Final    -0.1000000000000E+01   0.4000000000000E+01  -0.7500000000000E+00   0.7500000000000E+00
   BR=   0  BCR=   1  CR=   9  LS=   0  vfomin= 0.2098586473917E-07  vf0= 0.1718986654839E+00
-----------------------------------------------------------------------------------------------

73. BRODEN
   n= 4      N= 100     M= 100       iter=    20     nfunc=     203504        time=    135
                          lobnd              upbnd              lobndc             upbndc
Initial  -0.1000000000000E+02   0.1000000000000E+02  -0.5000000000000E+01   0.5000000000000E+01
Final    -0.6250000000000E+00   0.6250000000000E+00  -0.1250000000000E+00   0.1250000000000E+00
   BR=   4  BCR=   2  CR=   0  LS=   0  vfomin= 0.8582220162694E+05  vf0= 0.7926693336997E+07
-----------------------------------------------------------------------------------------------

74. HOSAKI
   n= 2      N= 500     M=  50       iter=     2     nfunc=      51661        time=      1
                          lobnd              upbnd              lobndc             upbndc
Initial   0.0000000000000E+00   0.2000000000000E+01   0.0000000000000E+00   0.1000000000000E+01
Final     0.0000000000000E+00   0.2000000000000E+01   0.0000000000000E+00   0.1000000000000E+01
   BR=   0  BCR=   0  CR=   1  LS=   0  vfomin=-0.2345811576100E+01  vf0=-0.7664155024405E+00
-----------------------------------------------------------------------------------------------

75. COSMIN
   n= 2      N=   5     M=  10       iter=    13     nfunc=        727        time=      0
                          lobnd              upbnd              lobndc             upbndc
Initial  -0.5000000000000E+03   0.5000000000000E+03  -0.1000000000000E+02   0.1000000000000E+02
Final    -0.5000000000000E+03   0.5000000000000E+03  -0.7812500000000E-01   0.7812500000000E-01
   BR=   0  BCR=   7  CR=  12  LS=   0  vfomin= 0.1544314623457E-09  vf0= 0.6500000000000E+02
-----------------------------------------------------------------------------------------------

76. BDQRTIC
   n= 16     N=  10     M=  10       iter=   320     nfunc=      37518        time=      3
                          lobnd              upbnd              lobndc             upbndc
Initial  -0.5000000000000E+04   0.5000000000000E+04  -0.5000000000000E+03   0.5000000000000E+03
Final    -0.4882812500000E+01   0.4882812500000E+01  -0.1192092895508E-03   0.1192092895508E-03
   BR=  10  BCR=  22  CR= 309  LS= 208  vfomin= 0.4229792213475E+02  vf0= 0.2712000000000E+04
-----------------------------------------------------------------------------------------------
```

```
     77. DIXON3DQ
     n= 10      N=  10     M=  50      iter=    55     nfunc=     28405        time=     1
                              lobnd               upbnd                lobndc               upbndc
Initial   0.0000000000000E+00   0.5000000000000E+03  -0.5000000000000E+02   0.5000000000000E+02
Final     0.0000000000000E+00   0.2500000000000E+03  -0.6103515625000E-02   0.6103515625000E-02
     BR=    1  BCR=  13   CR=  53   LS=   0  vfomin= 0.9099139103073E-01    vf0= 0.1300000000000E+02
---------------------------------------------------------------------------------------------------

     78. ENGVAL1
     n=  4      N= 100     M=  50      iter=    56     nfunc=    283757        time=     5
                              lobnd               upbnd                lobndc               upbndc
Initial  -0.5000000000000E+03   0.5000000000000E+03  -0.5000000000000E+02   0.5000000000000E+02
Final    -0.1562500000000E+02   0.1562500000000E+02  -0.2384185791016E-04   0.2384185791016E-04
     BR=    5  BCR=  21   CR=  49   LS=   0  vfomin= 0.2495604370956E+01    vf0= 0.1770000000000E+03
---------------------------------------------------------------------------------------------------

     79. EX-PEN
     n=  5      N=  10     M= 100      iter=    36     nfunc=     36596        time=     1
                              lobnd               upbnd                lobndc               upbndc
Initial  -0.5000000000000E+03   0.5000000000000E+03  -0.2000000000000E+03   0.2000000000000E+03
Final    -0.1562500000000E+02   0.1562500000000E+02  -0.1220703125000E-01   0.1220703125000E-01
     BR=    5  BCR=  14   CR=  30   LS=   0  vfomin= 0.1522038727647E+01    vf0= 0.3011562500000E+04
---------------------------------------------------------------------------------------------------

     80. BROYDENP
     n=  5      N= 100     M= 100      iter=    23     nfunc=    233749        time=     6
                              lobnd               upbnd                lobndc               upbndc
Initial  -0.5000000000000E+01   0.5000000000000E+01  -0.2000000000000E+01   0.2000000000000E+01
Final    -0.5000000000000E+01   0.5000000000000E+01  -0.7812500000000E-02   0.7812500000000E-02
     BR=    0  BCR=   8   CR=  22   LS=   0  vfomin= 0.5444996552378E+00    vf0= 0.3700000000000E+02
---------------------------------------------------------------------------------------------------

     81. TEO
     n=  2      N=1000     M=  10      iter=     6     nfunc=     69034        time=     3
                              lobnd               upbnd                lobndc               upbndc
Initial   0.0000000000000E+00   0.1000000000000E+01  -0.1000000000000E-02   0.1000000000000E-02
Final     0.0000000000000E+00   0.1000000000000E+01  -0.1000000000000E-02   0.1000000000000E-02
     BR=    0  BCR=   0   CR=   5   LS=   0  vfomin= 0.9000000000179E+00    vf0= 0.2402613308223E+01
---------------------------------------------------------------------------------------------------

     82. Coca
     n=  2      N=  10     M= 500      iter=     6     nfunc=     30081        time=     0
                              lobnd               upbnd                lobndc               upbndc
Initial  -0.5000000000000E+01   0.5000000000000E+01  -0.1000000000000E+00   0.1000000000000E+00
Final    -0.5000000000000E+01   0.5000000000000E+01  -0.1000000000000E+00   0.1000000000000E+00
     BR=    0  BCR=   0   CR=   5   LS=   0  vfomin= 0.7415391850332E-02    vf0= 0.1912500017166E+04
---------------------------------------------------------------------------------------------------

     83. NEC
     n=  2      N=  10     M=  50      iter=    11     nfunc=      5676        time=     0
                              lobnd               upbnd                lobndc               upbndc
Initial  -0.5000000000000E+01   0.5000000000000E+01  -0.1000000000000E+01   0.1000000000000E+01
Final    -0.5000000000000E+01   0.5000000000000E+01  -0.6250000000000E-01   0.6250000000000E-01
     BR=    0  BCR=   4   CR=  10   LS=   0  vfomin=-0.1249999999896E+01    vf0= 0.5000000000000E+01
---------------------------------------------------------------------------------------------------

     84. QPEXP
     n=  4      N= 100     M=  50      iter=     1     nfunc=      5212        time=     0
                              lobnd               upbnd                lobndc               upbndc
Initial  -0.5000000000000E+01   0.5000000000000E+01  -0.1000000000000E+01   0.1000000000000E+01
Final    -0.5000000000000E+01   0.5000000000000E+01  -0.1000000000000E+01   0.1000000000000E+01
     BR=    0  BCR=   0   CR=   0   LS=   0  vfomin= 0.8881666090321E-16    vf0= 0.7324861284383E-02
---------------------------------------------------------------------------------------------------

     85. NONDQUAR
     n=  8      N= 100     M=  50      iter=    10     nfunc=     51582        time=     2
                              lobnd               upbnd                lobndc               upbndc
Initial  -0.5000000000000E+02   0.5000000000000E+02  -0.1000000000000E+02   0.1000000000000E+02
Final    -0.1250000000000E+02   0.1250000000000E+02  -0.1000000000000E+02   0.1000000000000E+02
     BR=    2  BCR=   0   CR=   7   LS=   0  vfomin= 0.6822795617112E-10    vf0= 0.4900000000000E+03
---------------------------------------------------------------------------------------------------

     86. ARWHEAD
     n= 40      N=  50     M=  50      iter=    77     nfunc=    199656        time=    33
                              lobnd               upbnd                lobndc               upbndc
Initial  -0.5000000000000E+02   0.5000000000000E+02  -0.1000000000000E+02   0.1000000000000E+02
Final    -0.1250000000000E+02   0.1250000000000E+02  -0.1525878906250E-03   0.1525878906250E-03
     BR=    2  BCR=  16   CR=  74   LS=   0  vfomin= 0.1348798973067E-05    vf0= 0.1170000000000E+03
---------------------------------------------------------------------------------------------------

     87. CUBE
     n=  4      N= 500     M=  50      iter=  1735     nfunc=  44990111        time=   794
                              lobnd               upbnd                lobndc               upbndc
Initial  -0.5000000000000E+01   0.5000000000000E+01  -0.1000000000000E+01   0.1000000000000E+01
Final    -0.5000000000000E+01   0.5000000000000E+01  -0.1220703125000E-03   0.1220703125000E-03
     BR=    0  BCR=  13   CR=1734   LS=1695  vfomin= 0.9879035371252E-04    vf0= 0.1977237045971E+04
---------------------------------------------------------------------------------------------------
```

```
88. NONSCOMP
   n= 5      N= 100     M=  50      iter=   641      nfunc=    3329422        time=     71
                        lobnd                 upbnd                 lobndc                upbndc
Initial  -0.5000000000000E+01   0.5000000000000E+01  -0.1000000000000E+01   0.1000000000000E+01
Final    -0.5000000000000E+01   0.5000000000000E+01  -0.2441406250000E-03   0.2441406250000E-03
   BR=   0  BCR=  12  CR= 640  LS=   0  vfomin= 0.6114213167963E-05   vf0= 0.5800000000000E+03
------------------------------------------------------------------------------------------------

89. DENSCHNF
   n= 10     N=  5      M=  10      iter=    56      nfunc=       3196        time=      1
                        lobnd                 upbnd                 lobndc                upbndc
Initial  -0.5000000000000E+01   0.5000000000000E+01  -0.5000000000000E+01   0.5000000000000E+01
Final    -0.5000000000000E+01   0.5000000000000E+01  -0.1525878906250E-03   0.1525878906250E-03
   BR=   0  BCR=  15  CR=  54  LS=   0  vfomin= 0.1765232882554E-05   vf0= 0.2080000000000E+04
------------------------------------------------------------------------------------------------

90. BIGGSB1
   n= 16     N= 10      M=   5      iter=   744      nfunc=      49811        time=      4
                        lobnd                 upbnd                 lobndc                upbndc
Initial  -0.5000000000000E+01   0.5000000000000E+01  -0.1000000000000E+01   0.1000000000000E+01
Final    -0.5000000000000E+01   0.5000000000000E+01  -0.1220703125000E-03   0.1220703125000E-03
   BR=   0  BCR=  13  CR= 743  LS=   0  vfomin= 0.1791687822047E-05   vf0= 0.2000000000000E+01
------------------------------------------------------------------------------------------------

91. BORSEC6
   n= 30     N= 100     M=  50      iter=    14      nfunc=      72779        time=      7
                        lobnd                 upbnd                 lobndc                upbndc
Initial   0.0000000000000E+00   0.4000000000000E+01  -0.1000000000000E+01   0.1000000000000E+01
Final     0.0000000000000E+00   0.5000000000000E+00  -0.1000000000000E+01   0.1000000000000E+01
   BR=   3  BCR=   0  CR=  10  LS=   0  vfomin= 0.4367156132713E-06   vf0= 0.2168583873406E+12
------------------------------------------------------------------------------------------------

92. 3T-QUAD
   n= 2      N=1000     M=  90      iter=    21      nfunc=    1926877        time=      8
                        lobnd                 upbnd                 lobndc                upbndc
Initial  -0.5000000000000E+04   0.5000000000000E+04  -0.1000000000000E+04   0.1000000000000E+04
Final    -0.6103515625000E+00   0.6103515625000E+00  -0.1000000000000E+04   0.1000000000000E+04
   BR=  13  BCR=   0  CR=   6  LS=   0  vfomin= 0.3686757105806E+02   vf0= 0.1020162281649E+13
------------------------------------------------------------------------------------------------

93. MISHRA9
   n= 3      N=1000     M=  90      iter=    18      nfunc=    1650269        time=     12
                        lobnd                 upbnd                 lobndc                upbndc
Initial  -0.5000000000000E+03   0.5000000000000E+03  -0.1000000000000E+03   0.1000000000000E+03
Final    -0.9765625000000E+00   0.9765625000000E+00  -0.2500000000000E+02   0.2500000000000E+02
   BR=   9  BCR=   2  CR=   8  LS=   0  vfomin= 0.1575758116274E-06   vf0= 0.6976250929902E+27
------------------------------------------------------------------------------------------------

94. WAYBURN1
   n= 2      N=1000     M=  90      iter=    17      nfunc=    1559943        time=      7
                        lobnd                 upbnd                 lobndc                upbndc
Initial  -0.5000000000000E+03   0.5000000000000E+03  -0.1000000000000E+03   0.1000000000000E+03
Final    -0.9765625000000E+00   0.9765625000000E+00  -0.1000000000000E+03   0.1000000000000E+03
   BR=   9  BCR=   0  CR=   7  LS=   0  vfomin= 0.1775923522586E-08   vf0= 0.6288740000000E+06
------------------------------------------------------------------------------------------------

95. WAYBURN2
   n= 2      N=1000     M=  50      iter=    11      nfunc=     570063        time=      2
                        lobnd                 upbnd                 lobndc                upbndc
Initial  -0.5000000000000E+03   0.5000000000000E+03  -0.1000000000000E+03   0.1000000000000E+03
Final    -0.7812500000000E+01   0.7812500000000E+01  -0.1000000000000E+03   0.1000000000000E+03
   BR=   6  BCR=   0  CR=   4  LS=   3  vfomin= 0.1044643130973E-08   vf0= 0.1079181850156E+03
------------------------------------------------------------------------------------------------

96. DIX&PRI
   n= 10     N= 100     M=  50      iter=    64      nfunc=     331133        time=     14
                        lobnd                 upbnd                 lobndc                upbndc
Initial  -0.1000000000000E+02   0.1000000000000E+02  -0.5000000000000E+01   0.5000000000000E+01
Final    -0.5000000000000E+01   0.5000000000000E+01  -0.3051757812500E-03   0.3051757812500E-03
   BR=   1  BCR=  14  CR=  62  LS=   0  vfomin= 0.6666667207742E+00   vf0= 0.1093660000000E+06
------------------------------------------------------------------------------------------------

97. QING
   n= 15     N=  5      M= 500      iter=    81      nfunc=     203113        time=     13
                        lobnd                 upbnd                 lobndc                upbndc
Initial  -0.5000000000000E+03   0.5000000000000E+03  -0.5000000000000E+02   0.5000000000000E+02
Final    -0.1562500000000E+02   0.1562500000000E+02  -0.7629394531250E-03   0.7629394531250E-03
   BR=   5  BCR=  16  CR=  69  LS=   0  vfomin= 0.9277471785896E-06   vf0= 0.5200000000000E+03
------------------------------------------------------------------------------------------------
```

```
  98. QUAD-2V
   n=  2      N=1000     M=  50      iter=    12      nfunc=     618989          time=    3
                       lobnd               upbnd               lobndc              upbndc
Initial   -0.5000000000000E+03    0.5000000000000E+03   -0.5000000000000E+02    0.5000000000000E+02
Final     -0.3125000000000E+02    0.3125000000000E+02   -0.1250000000000E+02    0.1250000000000E+02
   BR=   4  BCR=   2   CR=   0  LS=   0  vfomin=-0.3873724182172E+04   vf0= 0.4388316000000E+05
------------------------------------------------------------------------------------------------

  99. RUMP
   n=  2      N=  10     M= 500      iter=    20      nfunc=     100338          time=    1
                       lobnd               upbnd               lobndc              upbndc
Initial    0.0000000000000E+00    0.2000000000000E-01    0.0000000000000E+00    0.1000000000000E-02
Final      0.0000000000000E+00    0.2000000000000E-01    0.0000000000000E+00    0.9765625000000E-06
   BR=   0  BCR=  10   CR=  19  LS=   9  vfomin=-0.6850076846409E+02   vf0= 0.4802237950000E+00
------------------------------------------------------------------------------------------------

  100. EX-CLIFF
   n=  4      N= 100     M=  50      iter=    60      nfunc=     311491          time=   12
                       lobnd               upbnd               lobndc              upbndc
Initial    0.0000000000000E+00    0.2000000000000E+00    0.0000000000000E+00    0.5000000000000E-01
Final      0.0000000000000E+00    0.2000000000000E+00    0.0000000000000E+00    0.5000000000000E-01
   BR=   0  BCR=   0   CR=  59  LS=   0  vfomin= 0.3995734044744E+00   vf0= 0.2000450000000E+01
------------------------------------------------------------------------------------------------

  101. NONDIA
   n= 10      N=  10     M=  50      iter=    83      nfunc=      42869          time=    2
                       lobnd               upbnd               lobndc              upbndc
Initial   -0.1000000000000E+03    0.1000000000000E+03   -0.1000000000000E+01    0.1000000000000E+01
Final     -0.6250000000000E+00    0.6250000000000E+01   -0.2441406250000E-03    0.2441406250000E-03
   BR=   4  BCR=  12   CR=  77  LS=   0  vfomin= 0.9898969787039E+00   vf0= 0.3604000000000E+04
------------------------------------------------------------------------------------------------

  102. EG2
   n=  4      N= 100     M=  50      iter=    17      nfunc=      86897          time=    3
                       lobnd               upbnd               lobndc              upbndc
Initial    0.0000000000000E+00    0.1000000000000E+03    0.0000000000000E+00    0.1000000000000E+02
Final      0.0000000000000E+00    0.1000000000000E+03    0.0000000000000E+00    0.4882812500000E-02
   BR=   0  BCR=  11   CR=  16  LS=   6  vfomin=-0.3499997998313E+01   vf0= 0.2945148446828E+01
------------------------------------------------------------------------------------------------

  103. LIARWHD
   n=  8      N= 100     M=  50      iter=    47      nfunc=     243103          time=    8
                       lobnd               upbnd               lobndc              upbndc
Initial   -0.1000000000000E+03    0.1000000000000E+03   -0.1000000000000E+02    0.1000000000000E+02
Final     -0.1250000000000E+02    0.1250000000000E+02   -0.122703125000E-02     0.122703125000E-02
   BR=   3  BCR=  13   CR=  42  LS=   0  vfomin= 0.1287394168882E-06   vf0= 0.4680000000000E+04
------------------------------------------------------------------------------------------------

  104. FULL-HES
   n=  4      N=  10     M= 100      iter=    19      nfunc=      19373          time=    2
                       lobnd               upbnd               lobndc              upbndc
Initial   -0.5000000000000E+03    0.5000000000000E+03   -0.1000000000000E+02    0.1000000000000E+02
Final     -0.9765625000000E+00    0.9765625000000E+00   -0.5000000000000E+01    0.5000000000000E+01
   BR=   9  BCR=   1   CR=   6  LS=   3  vfomin= 0.1212871287130E+02   vf0= 0.1660870312500E+07
------------------------------------------------------------------------------------------------

  105. NALSYS
   n=  2      N= 100     M= 500      iter=    39      nfunc=    1954856          time=   79
                       lobnd               upbnd               lobndc              upbndc
Initial   -0.1000000000000E+00    0.1000000000000E+00   -0.1000000000000E-01    0.1000000000000E-01
Final     -0.1000000000000E+00    0.1000000000000E+00   -0.1000000000000E-01    0.1000000000000E-01
   BR=   0  BCR=   0   CR=  38  LS=   7  vfomin= 0.2807057782276E-10   vf0= 0.1489929163060E-01
------------------------------------------------------------------------------------------------

  106. ENGVAL8
   n=  4      N= 100     M=  50      iter=    14      nfunc=      72067          time=    1
                       lobnd               upbnd               lobndc              upbndc
Initial    0.0000000000000E+00    0.1000000000000E+01   -0.1000000000000E+00    0.1000000000000E+00
Final      0.0000000000000E+00    0.1000000000000E+01   -0.1000000000000E+00    0.1000000000000E+00
   BR=   0  BCR=   0   CR=  13  LS=   1  vfomin=-0.3739004994563E+02   vf0= 0.2190000000000E+03
------------------------------------------------------------------------------------------------

  107. DIXMAANA
   n= 10      N=   5     M=  10      iter=     7      nfunc=        405          time=    0
                       lobnd               upbnd               lobndc              upbnd
Initial    0.0000000000000E+00    0.1000000000000E+01   -0.1000000000000E+00    0.1000000000000E+00
Final      0.0000000000000E+00    0.1000000000000E+01   -0.1000000000000E+00    0.1000000000000E+00
   BR=   0  BCR=   0   CR=   6  LS=   1  vfomin= 0.1000000012293E+01   vf0= 0.7400000000000E+02
------------------------------------------------------------------------------------------------

  108. DIXMAANB
   n= 10      N= 100     M=  50      iter=     7      nfunc=      36106          time=   10
                       lobnd               upbnd               lobndc              upbnd
Initial    0.0000000000000E+00    0.1000000000000E+01   -0.1000000000000E+00    0.1000000000000E+00
Final      0.0000000000000E+00    0.1000000000000E+01   -0.1000000000000E+00    0.1000000000000E+00
   BR=   0  BCR=   0   CR=   6  LS=   2  vfomin= 0.1000000011413E+01   vf0= 0.1380750000000E+03
------------------------------------------------------------------------------------------------
```

```
     109. DIXMAANC
     n=  5      N= 100     M=  50        iter=     6      nfunc=     30828            time=     4
                           lobnd                     upbnd                   lobndc                    upbndc
Initial    0.0000000000000E+00    0.1000000000000E+01   -0.1000000000000E+00    0.1000000000000E+00
Final      0.0000000000000E+00    0.1000000000000E+01   -0.1000000000000E+00    0.1000000000000E+00
   BR=   0  BCR=   0  CR=   5  LS=   0  vfomin= 0.1000000001937E+01   vf0= 0.1095000000000E+03
-------------------------------------------------------------------------------------------------------

     110. DIAGAUP1
     n=  5      N= 100     M=  50        iter=    44      nfunc=    228394            time=     4
                           lobnd                     upbnd                   lobndc                    upbndc
Initial    0.0000000000000E+00    0.1000000000000E+05   -0.1000000000000E+03    0.1000000000000E+03
Final      0.0000000000000E+00    0.9765625000000E+01   -0.6103515625000E-02    0.6103515625000E-02
   BR=  10  BCR=  14  CR=  33  LS=   0  vfomin= 0.7507319964675E-07   vf0= 0.4005000000000E+04
-------------------------------------------------------------------------------------------------------

     111. EG3-COS
     n= 10      N=  5      M=  10        iter=    49      nfunc=      2811            time=     1
                           lobnd                     upbnd                   lobndc                    upbndc
Initial    0.0000000000000E+00    0.4000000000000E+01   -0.1000000000000E+00    0.1000000000000E+00
Final      0.0000000000000E+00    0.4000000000000E+01   -0.7812500000000E-03    0.7812500000000E-03
   BR=   0  BCR=   7  CR=  48  LS=   0  vfomin=-0.9499999615584E+01   vf0= 0.2226137858737E+01
-------------------------------------------------------------------------------------------------------

     112. VARDIM
     n= 10      N=  5      M=  5         iter=    66      nfunc=      2161            time=     0
                           lobnd                     upbnd                   lobndc                    upbndc
Initial    0.0000000000000E+00    0.1000000000000E+02   -0.5000000000000E+00    0.5000000000000E+00
Final      0.0000000000000E+00    0.1250000000000E+01   -0.4882812500000E-03    0.4882812500000E-03
   BR=   3  BCR=  10  CR=  62  LS=   0  vfomin= 0.7730452422942E-07   vf0= 0.2198553145800E+07
-------------------------------------------------------------------------------------------------------

     113. NARRCONE
     n=  2      N= 80      M=  80        iter=    20      nfunc=    130469            time=     1
                           lobnd                     upbnd                   lobndc                    upbndc
Initial    0.0000000000000E+00    0.4000000000000E+01   -0.5000000000000E+00    0.5000000000000E+00
Final      0.0000000000000E+00    0.4000000000000E+01   -0.5000000000000E+00    0.5000000000000E+00
   BR=   0  BCR=   0  CR=  13  LS=   0  vfomin=-0.4999711763084E+00   vf0= 0.3280000000000E+03
-------------------------------------------------------------------------------------------------------

     114. ACKLEY
     n= 10      N= 100     M=  10        iter=    14      nfunc=     16323            time=     2
                           lobnd                     upbnd                   lobndc                    upbndc
Initial   -0.3276800000000E+02    0.3276800000000E+02   -0.1100000000000E+02    0.1100000000000E+02
Final     -0.3276800000000E+02    0.3276800000000E+02   -0.1100000000000E+02    0.1100000000000E+02
   BR=   0  BCR=   0  CR=  13  LS=   0  vfomin=-0.2718122988146E+01   vf0= 0.3875317250981E+01
-------------------------------------------------------------------------------------------------------

     115. WOLFEmod
     n=  2      N= 500     M= 400        iter=     2      nfunc=    401736            time=     5
                           lobnd                     upbnd                   lobndc                    upbndc
Initial   -0.3000000000000E+02    0.3000000000000E+02   -0.1110000000000E+02    0.1110000000000E+02
Final     -0.3000000000000E+02    0.3000000000000E+02   -0.1110000000000E+02    0.1110000000000E+02
   BR=   0  BCR=   0  CR=   1  LS=   0  vfomin=-0.3885618082985E+01   vf0= 0.8333333333333E+00
-------------------------------------------------------------------------------------------------------

     116. PEAKS
     n=  2      N= 50      M=  50        iter=   425      nfunc=   1087323            time=     4
                           lobnd                     upbnd                   lobndc                    upbndc
Initial   -0.2000000000000E+01    0.2000000000000E+01   -0.1000000000000E-01    0.1000000000000E-01
Final     -0.1000000000000E+01    0.1000000000000E+01   -0.1000000000000E-01    0.1000000000000E-01
   BR=   1  BCR=   0  CR= 161  LS=   0  vfomin=-0.6547921763872E+01   vf0= 0.4161726395744E-03
-------------------------------------------------------------------------------------------------------

     117. U18
     n=  3      N= 100     M= 100        iter=    79      nfunc=    800719            time=    11
                           lobnd                     upbnd                   lobndc                    upbndc
Initial   -0.2000000000000E+01    0.2000000000000E+01   -0.1000000000000E+00    0.1000000000000E+00
Final     -0.2000000000000E+01    0.2000000000000E+01   -0.1000000000000E+00    0.1000000000000E+00
   BR=   0  BCR=   0  CR=  78  LS=   1  vfomin= 0.2492301853743E+02   vf0= 0.1870000000000E+03
-------------------------------------------------------------------------------------------------------

     118. U23
     n=  2      N= 10      M= 100        iter=     1      nfunc=      1024            time=     0
                           lobnd                     upbnd                   lobndc                    upbndc
Initial    0.0000000000000E+00    0.5000000000000E+01   -0.1000000000000E+00    0.1000000000000E+00
Final      0.0000000000000E+00    0.5000000000000E+01   -0.1000000000000E+00    0.1000000000000E+00
   BR=   0  BCR=   0  CR=   0  LS=   0  vfomin=-0.1372874556465E-05   vf0= 0.1082682265893E+01
-------------------------------------------------------------------------------------------------------

     119. SUMSQUAR
     n=  5      N= 100     M=1000        iter=     6      nfunc=    601032            time=    16
                           lobnd                     upbnd                   lobndc                    upbndc
Initial   -0.1000000000000E+02    0.1000000000000E+02   -0.1100000000000E+01    0.1100000000000E+01
Final     -0.1000000000000E+02    0.1000000000000E+02   -0.1100000000000E+01    0.1100000000000E+01
   BR=   0  BCR=   0  CR=   5  LS=   1  vfomin= 0.6062571615241E-08   vf0= 0.3000000000000E+02
-------------------------------------------------------------------------------------------------------
```

```
  120. VARDIM8
    n= 10     N=  10    M= 100      iter=    30     nfunc=     30467       time=     2
                        lobnd              upbnd              lobndc             upbndc
Initial    0.0000000000000E+00   0.1000000000000E+01  -0.1000000000000E+00   0.1000000000000E+00
Final      0.0000000000000E+00   0.5000000000000E+00  -0.3125000000000E-02   0.3125000000000E-02
    BR=  1  BCR=   5  CR=  22  LS=   0  vfomin= 0.1462958086757E-07   vf0= 0.4830353687098E+13
-------------------------------------------------------------------------------------------

  121. MODULE
    n=  2     N=  10    M= 100      iter=    14     nfunc=     14192       time=     1
                        lobnd              upbnd              lobndc             upbndc
Initial   -0.1000000000000E+01   0.1000000000000E+01  -0.1000000000000E+00   0.1000000000000E+00
Final     -0.1000000000000E+01   0.1000000000000E+01  -0.1000000000000E+00   0.1000000000000E+00
    BR=  0  BCR=   0  CR=  13  LS=   0  vfomin= 0.7208138754557E-07   vf0= 0.2706705664732E+00
-------------------------------------------------------------------------------------------

  122. PEXP
    n=  2     N=  10    M= 100      iter=     5     nfunc=      5076       time=     0
                        lobnd              upbnd              lobndc             upbndc
Initial   -0.1000000000000E+01   0.1000000000000E+01  -0.1000000000000E+00   0.1000000000000E+00
Final     -0.1000000000000E+01   0.1000000000000E+01  -0.1000000000000E+00   0.1000000000000E+00
    BR=  0  BCR=   0  CR=   4  LS=   1  vfomin= 0.8484175001490E-12   vf0= 0.6064152299177E-01
-------------------------------------------------------------------------------------------

  123. COMB-EXP
    n=  2     N=  10    M= 100      iter=     3     nfunc=      3041       time=     0
                        lobnd              upbnd              lobndc             upbndc
Initial   -0.5000000000000E+01   0.5000000000000E+01  -0.1000000000000E+00   0.1000000000000E+00
Final     -0.5000000000000E+01   0.5000000000000E+01  -0.1000000000000E+00   0.1000000000000E+00
    BR=  0  BCR=   0  CR=   2  LS=   0  vfomin= 0.6760954994400E-07   vf0= 0.2706705817134E+00
-------------------------------------------------------------------------------------------

  124. QUADR1
    n=  2     N=  10    M= 100      iter=    31     nfunc=     31433       time=     0
                        lobnd              upbnd              lobndc             upbndc
Initial   -0.4000000000000E+01   0.4000000000000E+01  -0.1000000000000E+00   0.1000000000000E+00
Final     -0.4000000000000E+01   0.4000000000000E+01  -0.2500000000000E-01   0.2500000000000E-01
    BR=  0  BCR=   2  CR=  30  LS=   0  vfomin=-0.5999999998409E+01   vf0= 0.1990000000000E+04
-------------------------------------------------------------------------------------------

  125. QUADR2
    n=  2     N=  10    M=  50      iter=     8     nfunc=      4132       time=     0
                        lobnd              upbnd              lobndc             upbndc
Initial   -0.4000000000000E+01   0.4000000000000E+01  -0.1000000000000E+00   0.1000000000000E+00
Final     -0.4000000000000E+01   0.4000000000000E+01  -0.1000000000000E+00   0.1000000000000E+00
    BR=  0  BCR=   0  CR=   7  LS=   0  vfomin= 0.1000000000058E+02   vf0= 0.6000000000000E+02
-------------------------------------------------------------------------------------------

  126. SUM-EXP
    n= 50     N=   5    M=  50      iter=    21     nfunc=      5397       time=     3
                        lobnd              upbnd              lobndc             upbndc
Initial    0.1000000000000E+01   0.4000000000000E+01   0.1000000000000E+00   0.2000000000000E+01
Final      0.2500000000000E+00   0.1000000000000E+01   0.3906250000000E-03   0.7812500000000E-02
    BR=  2  BCR=   8  CR=  18  LS=   0  vfomin= 0.4900000200180E+02   vf0= 0.9841913092362E+03
-------------------------------------------------------------------------------------------

  127. CAMEL
    n=  2     N=  50    M=  50      iter=   679     nfunc=   1759498       time=    15
                        lobnd              upbnd              lobndc             upbndc
Initial   -0.5000000000000E+01   0.5000000000000E+01   0.0000000000000E+00   0.1000000000000E+00
Final     -0.1250000000000E+01   0.1250000000000E+01   0.0000000000000E+00   0.1862645149231E-08
    BR=  2  BCR=  29  CR=  38  LS=  43  vfomin=-0.9552512351121E+04   vf0= 0.2116666666667E+01
-------------------------------------------------------------------------------------------

  128. SP-MOD
    n= 10     N=  50    M=  50      iter=    17     nfunc=     43784       time=     1
                        lobnd              upbnd              lobndc             upbndc
Initial   -0.5000000000000E+01   0.5000000000000E+01   0.0000000000000E+00   0.1000000000000E+01
Final     -0.5000000000000E+01   0.5000000000000E+01   0.0000000000000E+00   0.1000000000000E+01
    BR=  0  BCR=   0  CR=  13  LS=   2  vfomin= 0.2934784437013E-03   vf0= 0.1100000000000E+02
-------------------------------------------------------------------------------------------

  129. TRECANNI
    n=  2     N=   5    M=   5      iter=    10     nfunc=       319       time=     0
                        lobnd              upbnd              lobndc             upbndc
Initial   -0.5000000000000E+01   0.5000000000000E+01   0.0000000000000E+00   0.1000000000000E+01
Final     -0.5000000000000E+01   0.5000000000000E+01   0.0000000000000E+00   0.1250000000000E+00
    BR=  0  BCR=   3  CR=   9  LS=   0  vfomin= 0.2024356506683E-09   vf0= 0.1000000000000E+02
-------------------------------------------------------------------------------------------

  130. ProV-MV
    n= 10     N=  10    M=  10      iter=    23     nfunc=      2677       time=     1
                        lobnd              upbnd              lobndc             upbndc
Initial   -0.5000000000000E+02   0.5000000000000E+02   0.0000000000000E+00   0.1000000000000E+01
Final     -0.5000000000000E+02   0.5000000000000E+02   0.0000000000000E+00   0.6250000000000E-01
    BR=  0  BCR=   4  CR=   0  LS=   8  vfomin= 0.2020482388347E+03   vf0= 0.2950000000000E+03
-------------------------------------------------------------------------------------------
```

```
      131. SCALQUAD
      n= 10      N=  10      M=  10       iter=    16      nfunc=      1859          time=      0
                             lobnd             upbnd             lobndc            upbndc
Initial  -0.5000000000000E+02   0.5000000000000E+02   0.0000000000000E+00   0.1000000000000E+01
Final    -0.5000000000000E+02   0.5000000000000E+02   0.0000000000000E+00   0.3906250000000E-02
      BR=   0  BCR=   8  CR=   0  LS=   0  vfomin= 0.4233492307100E+02   vf0= 0.5500000000000E+02
------------------------------------------------------------------------------------------------

      132. BRASOV
      n= 10      N=   5      M=   5       iter=    77      nfunc=      2468          time=      0
                             lobnd             upbnd             lobndc            upbndc
Initial  -0.5000000000000E+01   0.5000000000000E+01  -0.1000000000000E+01   0.1000000000000E+01
Final    -0.5000000000000E+01   0.5000000000000E+01  -0.4882812500000E-03   0.4882812500000E-03
      BR=   0  BCR=  11  CR=  17  LS=   0  vfomin= 0.6932289635176E+02   vf0= 0.1224000000000E+04
------------------------------------------------------------------------------------------------

      133. PROD-SUM
      n= 10      N=  50      M=  50       iter=     8      nfunc=     20611          time=      1
                             lobnd             upbnd             lobndc            upbndc
Initial  -0.5000000000000E+01   0.5000000000000E+01  -0.1000000000000E+01   0.1000000000000E+01
Final    -0.5000000000000E+01   0.5000000000000E+01  -0.1000000000000E+01   0.1000000000000E+01
      BR=   0  BCR=   0  CR=   7  LS=   0  vfomin= 0.1389960254143E-05   vf0= 0.5500000000000E+02
------------------------------------------------------------------------------------------------

      134. PRODPROD
      n= 10      N=   5      M=   5       iter=    10      nfunc=       334          time=      0
                             lobnd             upbnd             lobndc            upbndc
Initial  -0.5000000000000E+01   0.5000000000000E+01  -0.1000000000000E+01   0.1000000000000E+01
Final    -0.6250000000000E+01   0.6250000000000E+01  -0.2500000000000E+00   0.2500000000000E+00
      BR=   3  BCR=   2  CR=   6  LS=   0  vfomin= 0.2180037938679E-05   vf0= 0.1111111110100E+12
------------------------------------------------------------------------------------------------

      135. PS-COS
      n= 10      N=  50      M=  50       iter=    74      nfunc=    189030          time=     62
                             lobnd             upbnd             lobndc            upbndc
Initial  -0.5000000000000E+01   0.5000000000000E+01  -0.1000000000000E+01   0.1000000000000E+01
Final    -0.5000000000000E+01   0.5000000000000E+01  -0.4882812500000E-03   0.4882812500000E-03
      BR=   0  BCR=  11  CR=   5  LS=   0  vfomin=-0.6902392229727E+03   vf0=-0.4614893409920E+03
------------------------------------------------------------------------------------------------

      136. PP-COS
      n= 10      N=  10      M=  10       iter=   117      nfunc=     13636          time=      5
                             lobnd             upbnd             lobndc            upbndc
Initial  -0.5000000000000E+01   0.5000000000000E+01  -0.1000000000000E+01   0.1000000000000E+01
Final    -0.5000000000000E+01   0.5000000000000E+01  -0.2441406250000E-03   0.2441406250000E-03
      BR=   0  BCR=  12  CR=  33  LS=   0  vfomin=-0.6617719095465E+02   vf0=-0.3773267891917E+01
------------------------------------------------------------------------------------------------

      137. PS-SIN
      n= 10      N=  10      M=  10       iter=   122      nfunc=     13698          time=      4
                             lobnd             upbnd             lobndc            upbndc
Initial  -0.5000000000000E+01   0.5000000000000E+01  -0.1000000000000E+01   0.1000000000000E+01
Final    -0.5000000000000E+01   0.5000000000000E+01  -0.4882812500000E-03   0.4882812500000E-03
      BR=   0  BCR=  11  CR=   7  LS=  40  vfomin=-0.6131657360927E+03   vf0=-0.2992116109892E+03
------------------------------------------------------------------------------------------------

      138. PP-SIN
      n= 10      N=  10      M=  10       iter=   105      nfunc=     12237          time=      4
                             lobnd             upbnd             lobndc            upbndc
Initial  -0.5000000000000E+01   0.5000000000000E+01  -0.1000000000000E+01   0.1000000000000E+01
Final    -0.5000000000000E+01   0.5000000000000E+01  -0.2441406250000E-03   0.2441406250000E-03
      BR=   0  BCR=  12  CR=  39  LS=  37  vfomin=-0.6929157853667E+02   vf0=-0.3515404061961E+01
------------------------------------------------------------------------------------------------

      139. BOHA
      n=  2      N=  10      M=  10       iter=     8      nfunc=       903          time=      0
                             lobnd             upbnd             lobndc            upbndc
Initial  -0.5000000000000E+01   0.5000000000000E+01  -0.1000000000000E+01   0.1000000000000E+01
Final    -0.5000000000000E+01   0.5000000000000E+01  -0.1000000000000E+01   0.1000000000000E+01
      BR=   0  BCR=   0  CR=   7  LS=   0  vfomin= 0.5817829773491E-08   vf0= 0.3600000000013E+01
------------------------------------------------------------------------------------------------

      140. DECK-AAR
      n=  2      N=  10      M=  10       iter=   129      nfunc=     14723          time=      0
                             lobnd             upbnd             lobndc            upbndc
Initial  -0.5000000000000E+01   0.5000000000000E+01  -0.1000000000000E+01   0.1000000000000E+01
Final    -0.6250000000000E+00   0.6250000000000E+00  -0.1953125000000E-02   0.1953125000000E-02
      BR=   3  BCR=   9  CR=  11  LS=   0  vfomin=-0.2492930182626E+06   vf0= 0.9997000016000E+04
------------------------------------------------------------------------------------------------
```

Printed in the United States
by Baker & Taylor Publisher Services